The Art of Beekeeping

Bees and Humans:
A Partnership Through Time

Design and layout: David Miller
Illustrations and cover artwork: Szarka Fedor Guido
First edition: Gazsó Méhészet és Szakkönyvkiadó, Hungary, 2024

Published by Northern Bee Books in 2026
Scout Bottom Farm,
Mytholmroyd,
West Yorkshire
HX7 5JS (UK)
Tel: 01422 882751
Fax: 01422 886157
www.northernbeebooks.co.uk

ISBN 978-1-9192004-9-1

TIBOR WEINER SENNYEY

The Art of Beekeeping

Bees and Humans: A Partnership Through Time

Essays on Apiculture
Translated by Deborah Marshall

Northern Bee Books
2026

Contents

1

Bees of the Ancestors, or Prehistoric Honey Hunters and Bee Shamans

Neolithic man poked his nose out of the last ice age and began keeping bees. Many fine rock drawings from the Neolithic or New Stone Age show these very early beekeepers, or more precisely, honey hunters. The best known is the bee drawing found in the Cuevas de la Araña (Spiders' Cave) in Valencia, Spain, where two honey hunters can be seen climbing a tall tree, one at the top and the other at the bottom of a long rope ladder. When this drawing is reproduced, usually only the top part is visible, but the drawing actually shows a team of two honey hunters. The lower figure appears to be waiting, holding the rope at the base of the tree; the upper figure bears a pot in one hand, flailing at bees with the other. A swarm of disproportionally large bees surrounds him. A peculiarity of the drawing is that, if you look closely, you can see that the vessel which the honey hunter holds resembles a basket or, more likely, a smoker.

Prehistoric honey hunters.
Cuevas de la Araña (Spiders' Cave), Bicorp, Valencia, Spain.

Whether it's a smoker or a basket, it's likely that eight thousand to ten thousand years ago, he may have been trying to calm bees, much as do today's beekeepers. Humans have known and used fire for at least 100,000 years and have had plenty of time to observe the effects of fire and smoke on different animals. Bees suck their honeycombs - up to twice their body weight - in the smoke and prepare to fly out in swarms when the flames reach their homes. With a full honeycomb, bees do not attack or attack less, and it is uncommon for emerging swarms to sting. I wonder when humans first realized this. Probably at the very beginning of our relationship with bees.

Ancient cave drawing of a honey hunter and bees from the Cuevas de la Araña (Spiders' Cave) and their schematic drawing.

Incidentally, these cave drawings were made shortly after the last ice age, but even according to the best and most thorough researchers, such as Eva Crane *(The World History of Beekeeping and Honey Hunting, New York, 1999)*, as long as man has been around, he has sought and preferred sweet foods. And the sweetest and most energizing food in nature is honey, which our ancestors loved as much as we do. The cave paintings of ancient hunts may have been made for ritual purposes, but they show us honey hunts from the earliest times. Climbing ropes or leaning over the entrance to burrows, niches and caves, the honey hunters, at considerable risk, sought the coveted honey. In another equally amusing cave painting, we see not just two hunters, but a small group, many in comic postures. This drawing reminds me of the first spring flower honey spinning, which is just as much fun now in our own apiary. Here too, just as thousands of years ago, the bees sting some of the enthusiastic participants. Especially me.

John B. Free *(Bees and Mankind, London, 1982)* compares prehistoric honey hunters to tribal peoples, such as African tribes and Aboriginal peoples in Australia today, who still hunt honey using similar methods. Both aforementioned researchers, Crane and Free, offer many parallels between ancient honey hunters and contemporary hunter-gatherers who still preserve their original traditions. This suggests that ten thousand years ago we would have seen similar practices, tools and methods as used by the tiny groups of peoples whose prehistoric culture survives today. Of course, this is an assumption, but the schematic style of drawing and the mythic dimensions of cave drawings do indeed give us some notion of tribal life many thousands of years ago. Famous examples include the mystical elephant man with bees in the Royal Natal National Park and the drawing of the Zimbabwean honey collector (with a smoker in his hand), where the artist has even scored the honeycombs to see if they contained eggs, or as we Hungarian beekeepers say: spawned.

Group honey hunt from prehistoric times in another cave painting. Barranc Fondo, Valencia, Spain.

A schematic drawing of an elephant man surrounded by bees and the resulting African cave drawing

Among various tribal honey-hunting customs, the Bushman's belief that honey-gathering months are the most auspicious for marriage and love is particularly striking, and it is perhaps no coincidence that we are reminded of the word honeymoon. After all, in Hungarian folklore, spring and early summer, the time of honey gathering, have long been the time of love and courtship (as in the raising of the Maypole). It is worth noting that many tribal honey hunters eat the whole cut comb, wax and larvae included, and the piece of spawned honeycomb, which in itself abundant in protein, is a real delicacy. That this might be unappealing to us, but if we had to walk the savannahs for weeks before we could finally get some honey (and energy), we might have a different sense of taste. It wasn't just honey that was of value; in some places, traces of ancient wax residues as well as honey have been found. Thus, from the very beginning, beeswax was an important product for beekeepers, and it later became even more valuable in making candles to use as a source of light.

Honey hunter cave drawing and schematic drawing in Zimbabwe, Africa.

It is likely that prehistoric cave drawings were made less for documentation and more for sacred reasons, meaning that in most cases the hunt was drawn in advance, in the hope that through art, the success of the adventure could be encouraged. The sacred nature of these drawings has also been emphasized by Hilda M. Ransome's *The Sacred Bee in Ancient Times and Folklore, (London, 1937).*

As she asserts:

> "(...) the character and work of the bees remain the same, and we can never describe them as really domesticated animals; they often, as beekeepers know to their cost, display independence of character; it has been truly said that 'Bees never do anything invariably.'"

The way we put it is that no two years in beekeeping are ever the same. Over the millennia, it is we, our attitudes and our technology that have changed more than the bees themselves. And this consistency, despite everything, can always surprise the passionate and attentive beekeeper. It is perhaps worth recalling that the hive-frame beekeeping we know today dates back only a few hundred years; before that beekeeping was done in natural hives and stumps, in forests. Collecting honey then was little different from the ancient methods of honey hunting. Cave paintings from the Neolithic to the present day, to be found in a wide variety of places from Zimbabwe to India, show this vividly.

Even more fascinating in the search for the beginnings of the art of beekeeping is what we find in the so-called Fertile Crescent region, and especially in Anatolia. Thomas D. Seeley *(The Untold Story of the Honey Bee in the Wild. Princeton University. Press., 2019)* is not the only researcher to suggest that beekeeping and the first captive bees appeared immediately after the beginnings of agriculture. Indeed, Seeley hypothesizes that around the first major settlements, piles of empty, used clay pots were found and occupied by swarms of fledged bees. This, he believes, is how close observation of bees and their life cycles led to keeping swarms in horizontal clay pots and to harvesting honey without killing the swarms, may have begun. Fine examples of such practices appear later in the early high cultures – the Sumerians and ancient Egypt, where they developed very sophisticated methods of beekeeping and even practiced migration beekeeping.

For me, one of the most fascinating discoveries from the second half of the last century in what is now Turkey - next to the 12,000-year-old ruins of Göbekli Tepe - was Çatal Höyük. This magnificent Neolithic complex was first excavated by James Mellaart, and in this research at least fourteen different layers, different periods, have been differentiated. The settlement must have flourished 8800 years ago and become depopulated 7700-7500 years ago; in its heyday, it likely had at least 5000 inhabitants. They were already engaged in agriculture and animal husbandry, and they had rich architecture, material culture, as well as shrines and burial places, implying an intellectual and spiritual life. However, the most interesting wall paintings from the point of view of the art of beekeeping are those published in the book by Mellaart, who led the excavations *(Çatal Höyük, A Neolithic Town in Anatolia, McGraw Hill, New York, 1967)* as illustrations 41 and 42, which he describes as depicting two paintings with a honeycomb pattern painted on top of each other. The peculiarity of the pattern is that, if you look carefully, you can see larvae, insects and stylized flowers in the cells, which Mellaart suggests may have originally depicted the stages of the bee life cycle. These very early observations suggest a high level of knowledge of ancient beekeepers.

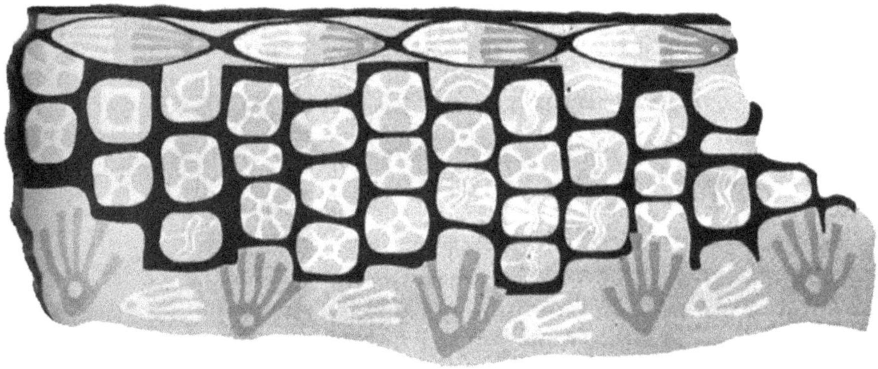

Photograph of the step-wall painting at the ancient archaeological site of Çatal-Höyük from Mellaart's book.

There are also drawings on the strange rock formations in the depths of the Sahara in Algeria in the labyrinthine mountain range of the Tassili n'Ajjer. This range is made of sandstone, which has created unique shapes through erosion. The traveler is confronted here with a lunar landscape, but one far richer in flora and fauna than the surrounding Sahara Desert. Thousands of years ago this was a paradise, and in the rock paintings found in the mountains, you can see people bathing in rivers with their rippling streams, countless animals that have now disappeared, vast herds, battles fought as senselessly between our common ancestors as we fight our wars today. Above all, you can see bird and insect life filling the air. But that glorious life has vanished beneath the grinding mills of millennia, and now only a hot desert wind howls through the dead shadows of dry rocks.

Terence McKenna writes in his book:

> "In the Tassili-n-Ajjer, rock paintings date from the late Neolithic to as recently as two thousand years ago. Here are the earliest known depictions of shamans with large numbers of grazing cattle. The shamans are dancing with

fists full of mushrooms and have mushrooms sprouting out of their bodies. In one instance they are shown running joyfully, surrounded by the geometric structures of their hallucinations. The pictorial evidence seems incontrovertible. *(Food of the Gods: The Search for the Original Tree of Knowledge – A Radical History of Plants, Drugs, and Human Evolution. New York, 1992.)*

I do not doubt what he wrote, except to note that what he describes as "a mushroom shaman wearing a beehive mask" is, to me, particularly interesting. We have a tendency, of course, to see ourselves in everything, which makes it difficult or impossible to see clearly. (I will write more about this later in an essay on *Bees and Man*.) What we can be sure of *here* and *now*, however, is that the depiction of mushrooms and the representation of bees in human form represents something best called anthropomorphization.

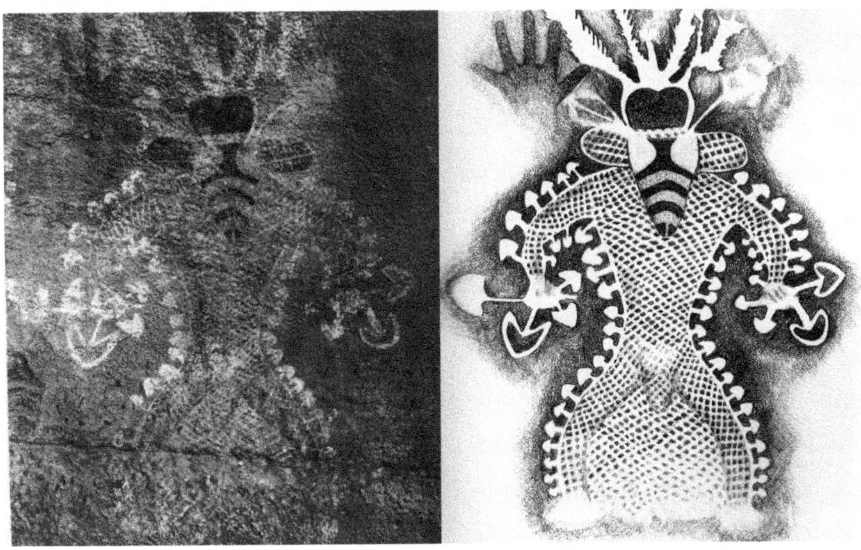

Depiction and schematic drawing of a bee shaman found among the rock drawings of Tassili n' Ajjer.

The best researcher of these petroglyphs was a French

photographer, Jean-Domonique Lajoux, who, if we are lucky enough to find his beautiful book *(Marveilles Du Tassili N'Ajjer, Paris, 1962)*, we can see that there is not just one but there are many more depictions of bee shamans in the petroglyphs! Lajoux points out that a similar beehive was found sixty kilometers away, and he believes that the now mysterious meaning of the figure could be "the local ghost" or "the representative of some kind of local myth". We don't know, because McKenna's name for the bee shaman is an afterthought, and we have no idea whether the images are of a *bee god* or a man in a bee suit and mask, performing some kind of ritual. But what is very interesting and common to all Stone Age drawings of bees and honey hunters is what Eva Crane has observed, namely that, unlike many other kinds of drawings, these bee-related images have not been drawn in the intervening millennia. Not in Africa, not in Europe, not in Asia. This can only mean that for thousands of years, generations of people have known that bee-related images and information are not to be rewritten or drawn; they are sacred and taboo, inviolable and fundamental.

Today, in the 21st century, we can see just how correct the ancients were. If bees and the knowledge and care of them are lost, then the whole of humanity will be lost, because bees are responsible for at least two-thirds of the pollination of flowers. So if there are no bees, there is no pollination, and there will be no crop yields. Without bees, humanity will die of starvation, will die out completely.

It is therefore an incomprehensible blunder and a murderous mistake in agriculture to use chemicals and technologies that are responsible for the mass extinction of bees and pollinators. This must stop immediately! Even primitive people knew that knowledge about bees was sacred and crucial to our survival. Agriculture, along with our villages and cities, must be made bee-friendly! We need to create bee-friendly gardens in our country, modelled on bird-friendly gardens! This should become a national movement!

Prehistoric man lived in close proximity to nature and was not alienated from it, but like us, he was also affected by climate change. In the rock paintings of the once prosperous and abundant in rivers region of Tassili n'Ajjer, we can see how, over thousands of years, cultures have changed, swimmers disappeared, beekeepers vanished, then nomadic migrants replaced endless herds, and finally the whole region dried up, became deserted as its former inhabitants abandoned it. Their drawings were no longer continued. Today, only a few Tuareg tribes still visit the area, the only scarce witnesses of the flourishing life and high culture are the rock paintings. Not so very far from this magical region, however, beekeeping soon blossomed again in the history of mankind, in the Nile Valley.

Daughters of the Sun, or the Bees and Beekeepers of Ancient Egypt

What is the difference between honey hunting and beekeeping? Even today, there are honey hunters who plunder nests to extract honey from the comb. But that is not so far from the practices of beekeepers who keep bees just for the honey and maximize their honey production without any concern for the bees' welfare. Actually, we should say that beekeeping only begins when bees are kept conscientiously, when the beekeeper's interest is as focused on the survival, health and development of the bees, as on the honey they produce. The earliest evidence of such educated, conscious beekeeping dates back 5,000 years to ancient Egypt, the time of the first dynasties, where we find the hieroglyph "bee" and the title "the sealer of the honey".

Bees were highly valued from the very beginning of the Ancient Egyptian Empire, and beekeepers were recognized in later times. It is true that in later ages, beekeepers were mainly priests, monks, teachers and hard-working, skilled craftsmen. In Egypt, the name of a ruler was preceded by the name of a bee-hieroglyph, together with a reed. The bee represented Lower Egypt, the fertile delta of the Nile, while the reed represented Upper Egypt. There are fascinating depictions, descriptions and drawings of bees, in which cylindrical clay tube hives, beekeepers, honey and beeswax from the ancient Egyptian millennia can be seen. Gene Kritsky (*The Tears of Re. Beekeeping in Ancient Egypt.* Oxford. 2015) has written extensively about this. He also includes a poem that preserves the belief in the divine origin of bees:

"The god Re wept, and the tears from his eyes Re?
fell to the ground
and became a bee.
The bee made honeycomb
And attended the flowers of every plant;
so beeswax came into being,
so honey came into being,
from the tears of Re."
(Papyrus Salt)

A Bee Hieroglyphic.

The god Ra (or Re) in ancient Egyptian mythology is the sun god, so, from very early times in the human belief system, bees were associated with creation, the sun and light.

Not by chance.

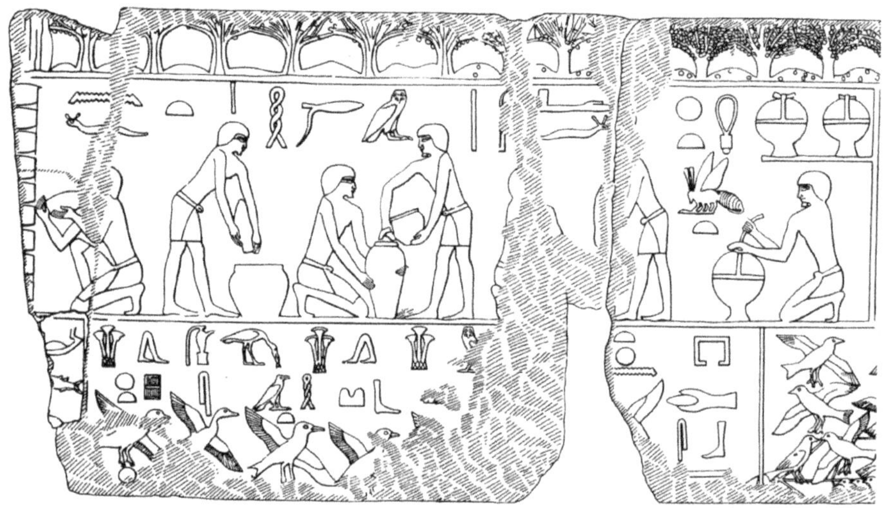

Ancient Egyptian beekeepers in a schematic drawing of a relief of the Sun Temple. The original site is the Sun Temple of Niuserre, Abu Ghorab, Egypt. The original relief is currently on display in the Egyptian Museum and Papyrus Collection, part of the Staatliche Museum zu Berlin.

Eva Crane writes that two mythical animals have accompanied humanity from the earliest creation stories: the snake and the bee. Both dwell in dark places and come out into the light, both have venom, but while the snake is a good or bad character in different mythologies and religions, the bee is a consistently positive character in all myths, religions and cultures. In ancient Egypt, as we have seen, they were "daughters of the sun". According to another myth, man himself was born from the tears and sweat of Ra, which can be interpreted as a direct relation between bee and man in the story of divine creation. Accordingly, perhaps because

Egypt was a high culture, to this day, bees have always played an important role in its history.

Detail of a wall painting of beekeepers preserved in the tomb of Rekhmire (c. 1479-1425 BC. Sheikh Abd el-Qurna.)

It was soon realized that bees produce not only honey but also valuable wax, and that they are most prized for their pollination. For example, in the 3500-year-old Sun Temple dedicated to the god Ra, we can see traces indicating that beekeeping was already a highly cultivated tradition. This temple was built by Niuserre, the sixth ruler of the 5th Dynasty of ancient Egypt, who reigned for 24-35 years in the 25th century BC.

A relief from this temple can be seen today in the Egyptian Museum and Papyrus Collection in Berlin. Even in fragments, the relief provides a wealth of valuable information and shows clearly that honey production of the time was a multi-stage process. On the left side of the picture, the cylindrical clay hives in which the bees were kept are stacked on top of each other, with a fragmentary figure kneeling in front of them with a strange vessel - more on this later. Then we see the honeycombs being placed in jars, pressed and poured, and finally sealed. Honey jars have been found that were sealed 3000 years ago, and the honey was still edible because honey – if sealed properly – does not spoil even after three thousand years

In other tombs (such as that of Rekhmire) and in surviving finds from various periods of the Egyptian Empire, beekeepers are shown processing honey, offering it to the pharaoh and to the gods in various temple ceremonies. We know that honey was also accepted as payment for taxes, so it had a great value, sometimes rivaling that of gold. Of course, honey's relative value has changed over the millennia just as it still does today.

Schematic drawing of the complete mural of beekeepers preserved in the tomb of Rekhmire.

Several papyri related to medicine, such as the Edwin Smith Papyrus, have also been found, and these may be based on an ancient text from around 3000-2500 BC, now attributed to Imhotep, the father of ancient Egyptian medicine. Another papyrus, the Ebers Papyrus from 1550 BC, also describes the medicinal use of honey. So, we know that honey was used for burns, wounds and even to treat and prevent eye diseases, especially infections. Now, in modern medicine, we also find so-called medicinal honey which is regularly used in winter honey as a cold remedy, although it might be better to eat it regularly as a preventive measure. Beeswax has also been used, not only in medicine, but also in art and for ritual purposes. The ancient Egyptians are still famous for their cosmetics, in which honey and especially beeswax were popular ingredients.

Honey was elevated to its pinnacle when it was offered as a sacrifice to the gods. Beekeepers themselves are described as having inhabited such positions as priests or beekeepers to a deity. Gene Kritsky suggests that the beekeeping community must also have been highly organized, as was the empire as a whole, which is perhaps why they were able to build such an amazing civilization with ensuing complexes of buildings, temples and pyramids, and to survive for so long. They built entire walls of their cylindrical horizontal hives made of clay from the Nile, and in some places in Egypt you can still see traditional beehives like this today. They used to load their bees onto boats and travel down the river, following the route of flowering plants. There are also records of different types of honey, such as the "white honey" of Lower Egypt and the "red honey" of the desert; of the two, the former was considered to be the purest. In his study, Manal B. Hammad, notes that the honey which first gushed out of the newly opened comb was considered "virgin honey" and the most valuable. The first pressed honey was used for normal consumption, and the harder pressed honey, which was left at the end, was used to make honey drinks and cakes. At least three different qualities were distinguished in honey production, and we know of at least two different types of honey, but there were probably more.

For me, the most interesting detail on the relief from the aforementioned Sun Temple relates to the figure kneeling on the left. There is some speculation that this beekeeper is holding a smoker, but as you can see in other depictions, the smoker – which was, of course, a favorite tool of ancient Egyptian beekeepers – looks different. Since the figure is fragmentary, and has thus captured the imagination of researchers, Crane and Kritsky also suggest that this is in fact a "singing out the queen", a now forgotten technique. In essence, when several queen bees are present, they start to make a tooting sound – this is what we call in Hungarian, the singing of the mothers – while the singing of queen bees who have not yet emerged from their eggs is called quacking. The singing queens look for each other in the hive by this sound, while the queens in the eggs who have not yet hatched are punctured. The sound indicates that the queen is ready to fight, wants to fight, but it can also indicate which queen should be supported by the new swarm. It appears that the beekeepers of ancient Egypt would listen for this sound, imitate it, and thus lure the queen and any subsequent swarm out of the hive, capturing the new family at once. If the image itself were not compelling enough evidence of this, there are several hieroglyphics on the relief describing the process, which represent "sing the hymn, fill, press, seal the honey". So, this is, as the English say, the "queen's piping", or as I would say "singing the mother out" from the hive. I note that the father of English beekeeping, Charles Butler, in his seminal 1609 beekeeping book *The Feminine Monarchie*, identifies the pitch of this sound (G#), but also that the first sound lasts for two seconds followed by several quarter-second sounds. Since Butler, in addition to beekeeping was also a musician, he later wrote a madrigal based on this sound. He published his score in the 1634 edition of his book and discussed at length how the sounds of bees could inspire human music. Perhaps this was echoed by his late compatriot and our contemporary, the London-based beekeeper and electronic composer Bioni Samp, who is trying to draw the attention of today's disconnected urbanites to the plight of bees with his particularly buzzing, noisy music inspired by the sounds made by the bees themselves.

I should note that, while it is a wonderful idea that the ancient Egyptian beekeeper in the relief could be "singing out" the queen bee, singing to the bees, this could be leading us astray, and it is possible that we are merely seeing a period smoker. In any case, in recent relevant research, Sophia Germanidou writes:

> "The earliest known representation of a smoker is found on a stone bas-relief from the temple of Ne-user-re, in Abu Ghorab, nowadays kept in the Egyptian Museum in Berlin (2400 BCE). An ovoid-shaped object, open at both sides, it depicts a beekeeper blowing smoke (?) through a hole towards the hives. However, the features of this object do not correspond to a smoker, despite what was initially argued. It is most probable that the depicted pot was used to induce the queen by 'emitting a breath or a little sound,' a practice known and applied by the Egyptian beekeepers for centuries. This practice survived in the traditional beekeeping of the Mediterranean – for instance it was known in Cyprus and in Greece. Support for this reading is provided by a depiction of a beekeeper scene in the famous tomb of Rekhmire in the West Bank of Luxor (c. 1450 BCE). Here the beekeeper holds a smoker shaped like a small deep bowl."

(Sophia Germanidou, Smoke and Bees: From Prehistoric to Traditional Smokers in Greece. New Approaches to the Archaeology of Beekeeping, Oxford: Archaeopress Publishing LTD, 2021, 19-29.)

Traditional Egyptian beekeeping flourished for thousands of years under the pharaohs, and after the last hieroglyphs were written and all pagan temples closed by order of the Christian Roman emperors, Egyptian traditional beekeeping has survived, in part and in traces, with its traditions, to this day.

It is no wonder that a very similar hive, made of clay and consisting of flattened cylinders, has been discovered by archaeologists in Tel Rehov, Israel, and is estimated to be at least three thousand years old. But the relationship of the Greco-Judeo-Christian culture to bees and honey, and therefore its technologies, was already different from that of ancient Egypt, but that is another story.

Bees and Canaan, “the Land Flowing with Milk and Honey”

The rich culture and art of beekeeping in ancient Egypt is best illustrated by the fact that the first relief depicting all the processes of beekeeping there dates back 4,500 years, as mentioned in the previous section. But the *Tale of Sinuhe*, written during the Middle Kingdom and considered one of the first surviving written literary works of mankind, recorded four thousand years ago, refers to “Canaan, filled with olive oil and honey”, to be precise:

> “It was a good land called Yaa.
> Figs grew there and grapes.
> It had more wine than water.
> Abundant was its honey, plentiful its oil.”
> *(Complaints of the Peasant. Ancient Egyptian short stories. Dobrovits A. (ed. [and trans.]), Magyar Helikon, Budapest, 1963.)*

For a long time, it was thought that the honey of the Holy Land might not have been honey at all, because although honey is frequently mentioned in the Scriptures, there is no reference to beekeeping. Moreover, some have argued that since Hebrew devašh (שבד) can mean both honey and other sweet syrups (e.g., date syrup) – so the archaeologist rightly asks in this excellent article:

> “In light of this, the question arises whether the ‘Canaan flowing with milk and honey’ was really flowing with honey?”
>
> *(Gabriella Nikoletta Kertész: Pollinators’ Month - Beekeeping, Pollination and their Symbolic Representations in the Ancient Near East.)*

But then adds:

"...the latest archaeological evidence suggests that there is a good chance."

Clay cylindrical hives very similar to the Egyptian ones have been discovered in Tel Rehov, Israel, and are estimated to be around three thousand years old. This sensational discovery is one of around 30 hives discovered between 1997 and 2012. The site, today's Tel Rehov (Arabic: Tell eṣ-Ṣarem), is located in a valley near fertile land and water sources, so perhaps it is no coincidence that it was here that intensive and conscious beekeeping was practiced. As one might rightly suspect:

"In the Bible, honey was not only a precious sweet, a special delicacy, but also a medicine (Proverbs 16:24), and thus often symbolized abundance and prosperity. It is no coincidence that the phrase 'Canaan flowing with honey and milk' is used to describe the Promised Land: God will bring his people into 'a land of honey' (II Kings 18:32). Be that as it may, the discovery of the Tel Rehov apiary has shed new light on what was previously thought."

(Miklós István Szabó, Israel as an Ancient Honey Power. Hitélet 60 (2022/5, 8.)

The Tel Rehov ancestral bee was found during the archaeological excavation of the city, surrounded by an urban environment, which means that the locals were not bothered by living among many hives of bees. Perhaps this was because this ancient beekeeper was also a good neighbor, perhaps he produced excellent honey, which smoothed over any possible unpleasantness or disputes. It's the same today if you are a beekeeper in a residential area, as I am.

A hive from Tel Rehov in Israel Museum, Jerusalem.

This is the world's first surviving beehive, about which we know due to a terrible fire that proved fatal for the beekeeper and the bees. These horizontal cylindrical hives with handles were made of clay mixed with straw and measure 80 cm in length, 40 cm in diameter, are 4 cm thick and have a 56-liter capacity. A small hole of 2-4 cm was left through which the bees could exit. Precisely because of the fire, the previously unfired clay pots were "fired" and have survived in surprisingly good condition. Even after three millennia, the remains of bees and honeycombs, beeswax, charred cereal grains and pollen grains were still found in them. The C-14 tests identified the age of the assemblage and concluded that this apiary was active exactly three thousand years ago in the northern kingdom of Israel in the late 10th and early 9th centuries BC. The honeybees were very likely brought here from Anatolia, in what is now Turkey, either via the Phoenician coast or by trade from Syria. During the fire, one of the walls collapsed, destroying most of the apiary. Amihai Mazar, a professor of archaeology at the Hebrew University of Jerusalem, and colleagues who led the excavation, wrote in the journal *Antiquity:*

> "The discovery of traces of industrial honey production significantly changes our understanding of the region's economy during that time. (...) Ancient beekeeping may have involved at least 75 hives, but they could number up to 200.

There were more than a million bees in the hive and had a potential annual yield of 500 kilograms of honey."

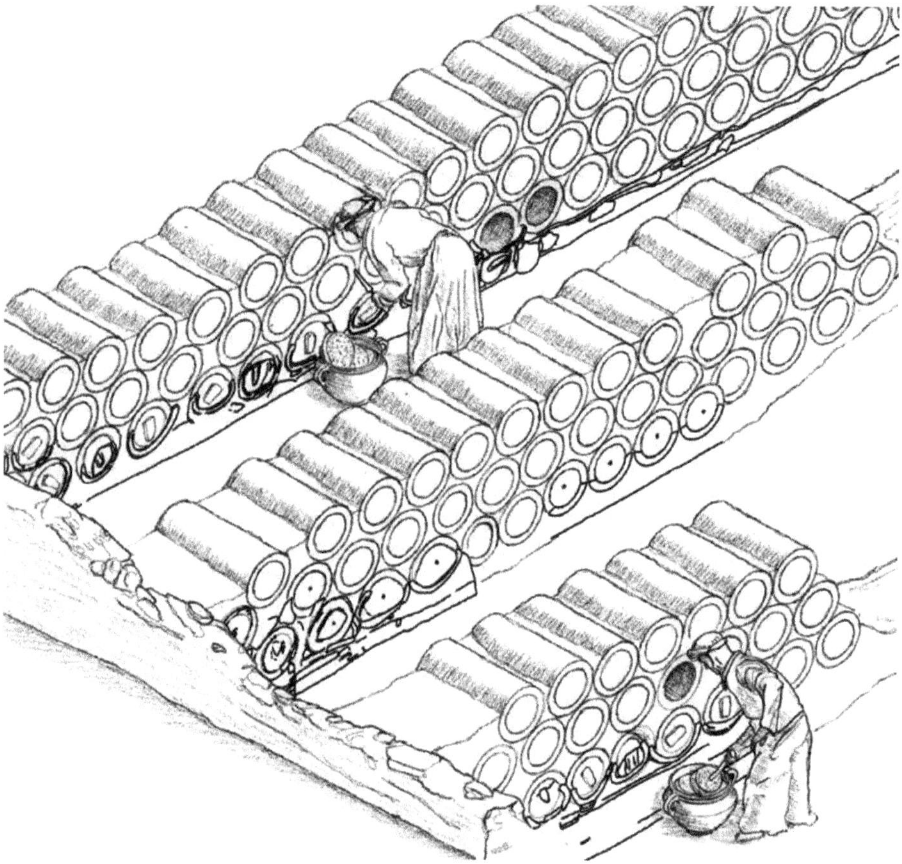

Reconstructed drawing of the Tel Rehov apiary.

These hives were made on the Egyptian model for Anatolian bees in Israel three thousand years ago.

What does this tell us?

It tells us that beekeeping has been a source of cultural and economic dialogue and interaction across borders and languages for three millennia. Obviously, this was not the only apiary in the Middle East; we just happened to find this one. It is almost certain that the ancient Jewish beekeepers were able to learn a lot from the Egyptians. It is also certain that if it was worth buying Anatolian bees, the beekeepers could even make long journeys to obtain, breed, and maintain them. Similarly, today it is worth investing in excellent queen bees and hives if we want to produce honey productively.

It would have been strange if the Jews had not known about beekeeping, since the Old Testament mentions honey fifty-five times, and Israel sixteen times as "Canaan flowing with milk and honey". Many Jewish holidays are still celebrated with honey cakes. According to a lesser-known tradition, Jewish children's lips are smeared with butter and honey at birth, perhaps to give them a taste of the nourishing richness and sweetness of life. Honey, honeycomb as a kiss, the bride's lips and love fulfilled, also appear in Song of Solomon, and elsewhere in Scriptures as wisdom incarnate. Thus, it is written:

> "My son, eat thou honey, because it is good; and the honeycomb, which is sweet to thy taste: So shall the knowledge of wisdom be unto thy soul: when thou hast found it, then there shall be a reward, and thy expectation shall not be cut off." *(Proverbs 24:13-14)*

Interesting parallels between the myths of different great traditions can be seen right here. Samson, known to all for his great strength carried in his long hair, his naive trust in a deceitful woman, his betrayal, his sad fate and his ultimate self-imposed prison was a great mythical hero of Judaism. Samson rose to prominence in the battles against the Philistines, just at the time when the apiary was about to burn down in Tel Rehov. Among Samson's many heroic deeds, one of the most interesting was when he defeated a lion, and later:

> "And after a time, he returned and turned aside to see the carcass of the lion: and behold, there was a

swarm of bees and honey in the carcass of the lion. And he took thereof in his hands, and went on eating, and came to his father and mother and he gave it to them, and they did eat." *(Judges 14:8-9)*

As with all myths, this must be understood metaphorically, but the interesting thing for me is that ancient Greece attributed almost the same story to Aristaeus, the son of the god Apollo and the nymph Cyrene. With the slight difference that Aristaeus was later worshipped not as a tragic hero but as the god of animal husbandry and beekeeping. This part of the myth seems to have been a kind of nomadic story. Given that bees are more likely to inhabit tree and holes in rock than the carcasses of dead animals, the myth of the swarm of bees emerging from the lion seems to me to be a powerful poetic image of the strength, significance, importance, and sacred essence of bees.

In early Christianity, honey appears primarily as the sustenance of St John the Baptist, who lived in the forest on locusts and honey – that is to say, he was a honey hunter, not a beekeeper – which interpreters of the scriptures present as symbols of judgment and prosperity; some even say that he who eats locusts and honey will be able to perceive good and evil, which is why he could recognize Jesus immediately. I don't know if merely eating locusts and honey would be sufficient to recognize Jesus; perhaps it is more useful to consider Jesus' teachings and to practice loving one's neighbor – but it is also an interesting parallel, even if a little distant, that when the Buddha became enlightened, the first gift he received from a monkey was a honeycomb. Buddha accepted it, so the monkey jumped from one branch to the next in joy until a branch broke underneath him and the poor thing died instantly. But the Buddha was not sad, for the monkey's good deed had enabled him to be born in a higher state of being, a human being, and thus to learn the Buddha's teachings and attain enlightenment. This story is echoed in countless wonderful works of art throughout Asia, and everywhere the monkey, honey in hand, is seen at the feet of the Buddha. (See *Robert L. Brown, Telling the Story in Art of the Monkey's Gift of Honey to the Buddha*.) Thus, the consumption of honey was associated with enlightened holy people by Christian and Buddhist believers.

The Emmaus Supper in Albrecht Dürer's engraving, where Christ ate fish and honey and then broke bread.

Honey and honeycomb have been highly valued delicacies and gifts in most all traditions. So much so, that when the disciples met in Emmaus and were visited by the risen Christ, this was reported:

> "And while they yet believed not for joy, and wondered, he said unto them: "Have ye here any meat?" And they gave him a piece of a broiled fish, and a honeycomb. And he took it, and did eat before them."
> *(Luke 24:41-43)*

Perhaps the blessing of honey, along with milk and water, may have been a survival of ancient rituals, such as the way the early Christians offered honey to the newly baptized. Later, the *"benedictio lactis et mellis"*, or blessing of milk and honey, was a ritual of the union of human and divine nature in man through baptism. It is true that honey still retains some of its significance at Christmastime, but unfortunately these old rituals with their deep meaning have faded from memory.

Yet it is precisely among the early Christian church fathers that we can find one of the first and most prominent connections between bees and poetry. Among the hymns of Clement of Alexandria (c. 150 AD - c. 215/220 AD, Greek Christian theologian, apologist and ecclesiastical writer), we find "To Christ the Educator", in which he beautifully compares believers to bees:

> "And following the example of the bees,
> As we gather the blossoms of the flowers,
> In the meadows, they swarm;
> And the reward of our toil,
> The sweet honey to their masters,
> And beyond the precious honey, the soft wax;

(I am such a servant of thing,
Like them, small and insignificant;
Gathering the flower of thy words,
Drunk with drink,
Blossoming into words to bless thee
My lips unbidden.)"
(Christ's Hymns of Clement of Alexandria.)

These hymns were traditionally sung before school and work. Whether or not we ourselves are practicing beekeepers, whether or not we are believers, we can recite the last lines of this poem as we leave for our hives:

"From poverty, Lord, keep me,
From riches keep me far away;
Rather I pray to thee,
That I may live in simplicity and modesty,
And happy shall be my arrival at my destination."

The Bee Film Festival

To get my readers in the mood for the rest of my essays, let's take a wide-angle detour and stage in our imagination *The Bee Film Festival.* We would divide the film festival into three categories. In the first block, during the morning hours, we would screen films especially for children. Here we would show an adaptation of the 1975 Austrian-Japanese and originally West German series of the now classic tale *Maya, the Bee.* More recently, 3-D versions of this tale have been made, but unfortunately, they are less convincing. The original animation with Hungarian dubbing was, however, quite hypnotic when I was a child, especially with the theme song by Karel Gott, whose Hungarian version is also - how shall I put it... – "wonderful».

The other „great film" that should definitely be shown is *Winnie the Pooh* from 1977, if possible with the original dubbing. Both were defining films for me as a child, especially *Winnie the Pooh*, since I was already a big fan of honey. Some of you may know that *Winnie the Pooh* became *Micimackó* in Hungarian from the English original because the genius poet Frigyes Karinthy had a linguistically brilliant but rather crazy sister called Emilia (Mici), who was the first to translate the work, which the poet then adapted. According to Márton Karinthy's recollection, "Aunt Mici was, nonetheless, completely mad. Once we were sitting with very fine people, professors, writers. At one point, Aunt Mici knelt down quite unexpectedly, threw herself on her knees, and cried out, 'Just one loaf of bread and butter for the poor widow!'" Not a single picture of her survives, but Karinthy preserved her memory in his translation of *Winnie the Pooh*. Critics say that despite having almost nothing to do with the original, it is brilliantly done as a philosophical work rather than just a simple children's book based on wordplay. There are several scenes in this story that I still love, such as Pooh's morning exercise, his being trapped in Bunny's hole, and especially the "pensive" moments. But the main charm of this tale is that the characters are archetypal characters, so it is very easy to identify with them, or to recognize yourself and your surroundings in a character. *Winnie the Pooh* is a fundamental work of world literature in which honey plays a key role, so we would definitely like to see a film version of it at the Bee Film Festival.

From Zsolt Magyar Poroszka's film The Acacia and Its Honey – The Story of the "Most Hungarian" Tree.

The children's section, also highly recommended for adults, would be followed by educational presentations about pollinators and beekeepers, and then finally the documentary block. Here, if possible, I would start with the films of Zsolt Magyar Poroszka, with *The Wonderful World of Bees*, which is also enjoyable for children, and *A Year in the Apiary*, and finally *Bee Pastures*. I really like Poroszka's films, I think they are incredibly well photographed and impeccably narrated, and a thousand times better than the sensationalist American bee documentaries I have seen here and there on educational channels.

That said, I would still screen *Vanishing of the Bees*, a 2009 film in which the horrors of industrial beekeeping and the impact of 21st-century disasters and human activity on bees are well documented, including the details and tragic consequences of agricultural pesticides, the Asian Varroa destructor, hive abandonment and similar horrors and their tragic consequences. We could even show a documentary on Africanized bees, but unfortunately the ones I have seen on this subject so far have not been exhaustive enough in their discussion of what caused this phenomenon in America. What? Human stupidity.

It would be really instructive and important to show the documentary about Brother Adam (*The Monk and the Honeybee*) with Hungarian subtitles; we could learn a lot from it. Maybe afterwards there could be a discussion about his book *Beekeeping at Buckfast Abbey*), which was recently published in Hungarian (translated by Imre Gazsó, 2021). Also in the documentary section, we should show the Hungarian black and white short film *The Migratory Beekeepers* (1955) and Szilard Dexler's short film about György Buchinger, *The Beekeeper's Refuge*. Both are very interesting and show the history of beekeeping in Hungary as it evolved over the last half century. We could probably find a few more of these short films, but we must now move on to feature films, which would make up the evening programs.

A still from Zsolt Magyar Poroszka's film A Year in the Apiary.

One of the most famous and interesting bee films in the world is *The Secret Life of Bees* from 2008, based on Sue Monk Kidd's 2001 novel of the same name. Exceptionally well adapted and extremely sensitive, the film tells the story of a troubled, world-weary little girl, Lily, and the plight of African Americans in the South in 1964. The three sisters who take Lily in are beekeepers, and as we delve deeper into their story, we understand why it is a "cult" film in America and why it is still being adapted into stage musicals. Queen Latifah is a wonderful beekeeper in it, and one of the most beautiful scenes is when she teaches Lily to be a beekeeper. If only because when we talk about beekeepers, a lot of people (almost everyone I've spoken to) think beekeepers as men, but there are many wonderful female beekeepers!

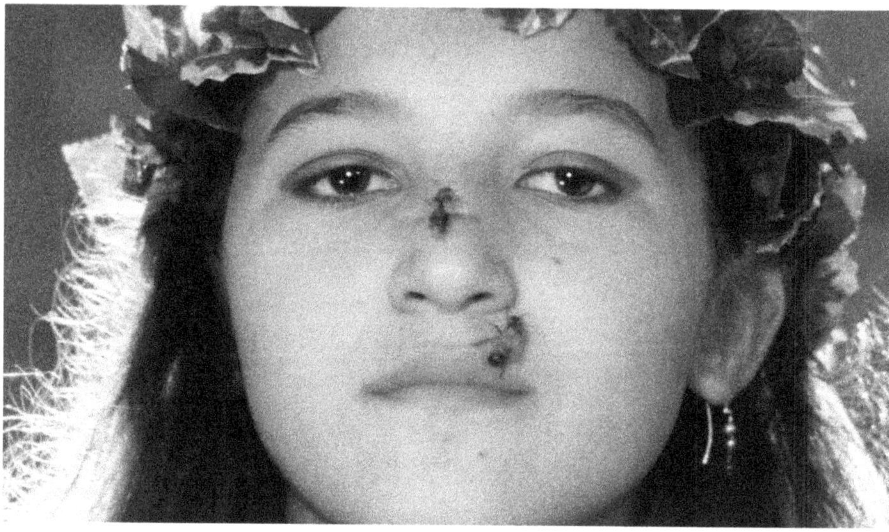

A frame from the film Le meraviglie (The Wonders)

Also a feature film, and perhaps most likely to appeal to young audiences, although I confess that I quite simply love it too - and not just because Monica Bellucci is in it – is the Italian *Le meraviglie (The Wonders)* from 2014. It is no coincidence that this film received many positive reviews at the Cannes Film Festival. Our protagonist is an adolescent girl whose father tries to support his family in the countryside by keeping bees, but this comes with a lot of tension and, of course, hard work. *The Wonders* is both about a little girl's coming of age and the difficulties of family beekeeping in Italy. Beautiful.

Also exploring a sensitive subject is the 2018 romantic drama *Tell It to the Bees*, which screened at the Toronto Film Festival and takes us to 1950s Scotland. I would also show the recent Turkish film *Kovan (Keeping the Bees)*, released in 2019, which takes place after the death of a beekeeper mother and tells the story of how her "urbanized" daughter takes over the family beekeeping business, facing difficulties and conflicts (and bears!) in the apiary on the border of a national park.

After these four great films, I would like to hold a panel discussion on why bees appear in these works of art in the way they do: in close connection with intimacy, with feminine forces, but also as sources of beauty and health – not only physical but also spiritual – as beings that help us to accept, to live and to mourn.

A still from the Bhutanese-Hungarian documentary film The Bees of Shambala, directed by Suraj Bhattarai (2022).

The fifth feature film, my personal favorite, which is difficult to talk about without bias, is *Mr. Holmes* from 2015. Set in the beautiful English countryside in 1947, Sherlock Holmes (played brilliantly by Ian McKellen), now an old man struggling with dementia, is a beekeeper. He is teaching his housekeeper's son the basics of beekeeping, while trying to put together the pieces of the puzzle of one last great mystery. Sir Arthur Conan Doyle really did lay the foundations for the story of this film in his short stories, but I'll write about that another time. The point is that the acting is quite extraordinary, the storytelling superb, the beekeeping parts are beautiful, and I must admit that when I first saw it, I had the idea that when I retired, I might be at least a little bit like Sherlock Holmes, whom I had always considered my role model. Anyway, there is a strange similarity between detective stories and beekeeping. You look for clues and signs and try to decipher what is happening. The more experience one has, the more skillful and attentive one is, the more quickly and accurately one gets to the heart of the matter. But what happens when the genius beekeeper-detective's mind is overwhelmed by a great veil of forgetfulness? Will he remember what he forgot? This film is about the great drama of old age, presented very subtly and compassionately, which is why I love it.

Another still from the Bhutanese–Hungarian documentary film The Bees of Shambala, directed by Suraj Bhattarai (2022).

At the imaginary festival's conclusion, the last two films would again be documentaries, although they are so strong that they are better than many feature films. *Der Imker (The Beekeeper),* directed by Mano Khalil, is a German film that tells the story of an elderly Kurdish beekeeper who fled to Switzerland. In this quite harrowing work of art, we not only see that the much-debated refugee issue in Europe is much more complex than it appears on the news, but we also discover that once a true beekeeper, always a beekeeper, even far from his war-torn homeland.

I should mention here that, while this book was being edited (in 2023), a great Spanish film was released, *20.000 especies de Abejas (20,000 Species of Bees)*, which is a truly captivating, sensitive and extremely beautiful work of art, and also touches on the secret of healing with bee stings and bee venom, which will be discussed later. Also sure to be a hit with audiences is the American action film *The Beekeeper*, which will be released in 2024, in which a beekeeper mercilessly slaughters all the bad guys. The thought of presenting such a commercial action film made me laugh out loud, but it does accurately portray what heroes beekeepers are and the hard work of beekeeping, so we should probably screen it.

A frame of the film Honeyland

I'd like to close the *Bee Film Festival* with a film that was marketed in English as *Honeyland* but was originally called *Medena zemja* (meaning *Honey Queen*). This is a very peculiar film, which follows the life of one of the last traditional beekeepers, a Macedonian woman, as she whispers mysterious incantations to her bees, just like the ancient Egyptian beekeepers. One of the important and recurring scenes in the film – and this is why I have now fictionally produced the *Bee Film Festival* – portrays the heroine visiting beehives in the cliffs, just as did the ancient, traditional Mediterranean beekeepers. We can follow exactly how she works with these stone beehives. It is likely that such hives date back to the most ancient beekeeping traditions. Traces of these beehives can also be found in Hungary, but their former use in this country is only preserved in folklore where they are called beehive stones. But since their original purpose has been shrouded in oblivion, they are associated with all sorts of mystical ideas. In my next essay I will write about these mystical beehive stones.

Beehive stone from the film Honeyland

The Mysterious Beehive Stones

In my previous essay, I would have closed the imaginary *Bee Film Festival* with a film originally titled *Medena zemja* (meaning *Honey Queen*). An important and recurring scene in the film is that our protagonist, just like traditional Mediterranean beekeepers, visits beehives carved into cliffs, often in very inaccessible hillsides. We can follow exactly how she works with these stone beehives which, we suspect, are part of a very ancient beekeeping tradition. In Hungary, too, we find traces of this ancient tradition, but their former use in Hungary is only preserved in folk memory, and they are therefore called beehive stones). In our country, like almost everything else, there are several seemingly contradictory explanations for beehive stones. I first came across these stones in Lajos Szabadfalvi's authoritative work on the history of Hungarian beekeeping, which contains many sources. Szabadfalvi puts it this way:

> "Before I start to outline the post-Hungarian Conquest beekeeping, I must mention a very interesting beekeeping-related monument, the so-called hive stones. To the north of the Great Plain, on the borders of Pest (Érd and Dunakeszi), Heves and Borsod counties (mainly in the southern part of the Bükk Mountains), there are abruptly rising cliffs, rock walls, with niches the size of beehives carved into the sides. Some of them have edges that are carved in such a way as to suggest that they have been sealed with planks. No concrete evidence of beekeeping has been found, but it is assumed that these rock shelters were once used to house bee

colonies, mainly on the basis of Mediterranean analogies. The Hungarian beehive stones were processed by Andor Saád and Péter Mihály."

(József Szabadfalvi: The Past of Hungarian Beekeeping. Debrecen, 1992, 12.)

Today, there is quite an extensive literature and history of research on beehive stones, but everyone mentions the first writer who dealt with this phenomenon, Gyula Bartalos. In 1891, Bartalos asked one of the central questions about beehive stones:

> "The question is what was enclosed in these hundreds and once thousands upon thousands of niches? (...) On the basis of the indications mentioned so far, we are perhaps entitled to form an opinion, and I believe that I am not going too far when I say that these niches are ash containers."
>
> *(Gyula Bartalos: "Beehive stones" and caves in Eger and environs. Eger: Archaeological Bulletin, January 1891.)*

So, according to Bartalos, the beehive stones were originally urns and had nothing to do with beekeeping. Perhaps closest to Bartalos' idea is the view of one of the best-known hive researchers, Csaba Baráz, who says

> "If we approach the question of the existence of niches, holes, bowl-shaped depressions, channels carved into the surface of the rocks from the point of view of the peculiarities of this 'pagan' religion, of the beliefs of Eastern societies in general - and for this attempt we will refer to the Hungarian Mythology of Arnold Ipolyi (IPOLYI ARNOLD, 1854), then the holes, bowl-like recesses, and stonework with drains carved into the tops of the beehive stones of the Törökasztal [rock formations called Turkish table] in Sirok, the Nyerges [saddled hill] in Eger, the Bányaél [mine edge] in Demjen, the Hegyes-kő-tető [stone roof] and the other beehive stones of the Bükkalja

area will bring to life the cultic rituals of the past. "
(Csaba Baráz. Barak Bükk in the Bükk Mountains. Debrecen, 1999.)

Csaba Baráz excludes the possibility of beekeeping and definitely interprets the beehive stones as the sites of pagan, ancient Hun and/or ancient Hungarian ceremonies. The Hungarian Wikipedia page on beehive stones is mostly in agreement with Baráz's writings, and states:

> "Despite the scarce archaeological and ethnographic data, the view has become accepted and known to the public that the beehive stones are the relics of medieval forest beekeeping and that this form of beekeeping was either introduced by the Kabars, who joined the Hungarians in the conquest period, or by a group of Thracian-Illyrian people, the Agrianes, who migrated from the Balkans in earlier times.
>
> "However, there are some doubts about the use of hive niches for beekeeping. One might wonder about the variation in location of the niches, as well as the presence of niches carved close to the ground or inaccessible, in dark, cool rocky crevices or in the walls of gullies. But narrow and shallow niches, or those that protrude, may preclude beekeeping. From the 11th century onwards, written records attest to the existence of beekeeping, with documents mentioning forest beekeepers and bee hunters, but no mention of rock beekeeping. Oral tradition is divided between cultic and economic use. Hence, the mystery of the beehive stones has not yet been solved."

While reading through the available literature, I decided to walk to the Nyerges Hill, where the only beehive stone in Szentendre has been found, next to our small apiary (Szentendre, Kada Peak). I approached the side of Nyerges from Pismány on a cool afternoon in late winter. The very imposing south-west-facing mountainside below the Asztal Stone, dotted with numerous natural rock cavities, offers a magnificent view of the forests below, the Kőhegy and the entire Pilis. After a little adventure, I found the Hive Stone, and while admiring the sunset, I thought long and hard about its purpose. After all, if these were what folklore says they are, they are the oldest beekeeping relics in Hungary. How interesting, I mused, that one of them is right here in my neighborhood.

The beehive stone in Szentendre.

I was concerned about this beehive stone issue, and I continued my research. Szabadfalvi mentioned two authors, Saád and Mihály. Saád wrote this in 1972:

> "Supplementing the beehive-stone data with that from my further investigations, I turned to the internationally

renowned and distinguished author, Zoltán Pál Őrösi, whose reply was in complete agreement with Vilkuna's and who also gave me valuable guidance in regard to the appropriate literature. He referred, among other things, to a method of beekeeping observed by Nachtsheim in Macedonia around Lake Katlanovo, south of Skopje, which also explains the purpose of our niches. Here, cavities formed in loess walls were covered with boards. Some of the bee families who had made their homes in the hives were relocated to the village. According to Pál Zoltán Őrösi, even the rarely occurring north-facing niches does not mean beekeeping did not happen there."

Perhaps that is why the Macedonian film mentioned above is an interesting analogy, in which we see such beekeeping. It is worth noting here that there is ample literature on the keeping of bees in rock cavities in the Mediterranean. In one of the studies on this subject it is written:

> 'The first written reference to beehive stones comes from the abbot Alberto Fortis, who described his experiences while passing through Dalmatia in 1774. On the island of Brac, he came across several hives made of stone slabs. The uppermost stone slab served as a roof, it was completely removable, and additional stone slabs were put together to protect against strong winds."
> *(Georgios Mavrofridis: Stone Beehives on the Islands of the Eastern Mediterranean, 126.)*

The author then lists the typical beehive stones of Corfu, Cephalonia, Rhodes and other islands.

Saád also suggests a possible link between the beehive stones and the nearby cave dwellings, dating their origin to between the 11th and 15th centuries. The question of ethnicity is a difficult problem, but he leans towards the Kabars, Kazars or Slavs who arrived with the conquerors, rather than toward Hun-Magyar origins. He rejects ritual purposes, referring mainly to the 1960 excavation at Cherepváralja, where no urn remains were found, and takes Örösi's comments where he advocates the presence of beekeeping very seriously. Thus, he writes about the semi-domestication of bees in the case of hive stones:

> "It is also possible that this method of beekeeping is an invention of the inhabitants of the tufa villages of the Rhyolite. However, it has its origins in the past, because this method of beekeeping can be described as 'semi-domestication', since it is a middle way between bees that are hidden in rocky outcrops and hollow trees and brought to the dwellings as pets."

He closes by saying:

> "I explain the purpose of the hives by the use of beekeeping adapted to the landscape. This latter view is confirmed by the fact that I myself found the remains of several rows of honeycomb on the vaults of a very intact niche in Szomolya, well protected by foliage and difficult to access, which indicate the presence of honeybees."
> *(Andor Saád: Data for determining the age and function of the hive stones. Annual of the Herman Otto Museum X, 108.)*

All of this strongly contradicts the notions most commonly held by Csaba Baráz and his followers that the hive stones would have served some ritual purpose.

Beehive stones from Szomolya.

Undoubtedly the most thorough researcher of the hive stones was Péter Mihály, one of whose most important writings on the subject is still in manuscript form in the manuscript archives of the Hungarian National Museum where I read it. The manuscript goes through the beehive stones in Hungary with respectable thoroughness, just like the high-quality Téka publication entitled *Sacral Stone Monuments - Hive Stones in Hungary*. But as the title of the latter shows, this publication considers the debate settled, i.e., that the hive stones were not used for beekeeping but for some sacral purpose.

The manuscript work of Peter Mihály, however, rejects this interpretation. He analyses at length the volume of the cavities, the remains of the carved frames and the occasional bee holes, and concludes his manuscript in this way:

> "In any case, one can wonder about the frequency of use of the niches. In the case of beekeeping, contact with the hives had to be made and the cover removed twice a year, i.e., when the swarm was being introduced and the beeswax was being extracted. For the hypothesis of burial urns, the niche remained unchanged for years after the initial placement of the urn and covering with the protective lid. In the case of the presentation of an offering, the frequency of the ceremony (unknown to us) determined the number of approaches. In several of the niches, the wedge holes used to affix the cover plate had worn out, so a new one was drilled or a cut after the previous frame had worn out. These observations, if true, point to frequent use of the niches, or rather to years of activity. Theoretically, therefore, this rules out the possibility of the burial of urns, which has so far only been refuted by the fact that no prehistoric urn fragments were found under the niches during the 1960 excavation at Cserépváralja."
> *(Péter Mihály, Hungarian National Museum manuscript archive, 1990, 63.)*

The largest and most exciting occurrences of hive stones are near Eger, in Szomolya and Cserépváralja, and are the most beautiful and spectacular finds. I decided it would be best to view them before committing to an opinion.

We started with the Hive Stone Study Trail of Szomolya, which has been extensively developed, and it should be added, thanks to enthusiastic hive stone research, the hive stones in Hungary have been protected since 2014. The trail is very popular with hikers, the large metal steps are very helpful for the less experienced, but I must admit that I found it a bit alien. The hive stones of Szomolya are in a sunny valley, mostly facing south-east, they are really spectacular, really beautiful. One of the plaques reads:

"Cone VII is reminiscent of the tufa towers of Cserépváralja. It has 3 niches, of which the longitudinal axis of No. 1 is so oblique as to call into question the possibility of keeping bees."

- and the like. The hive stone path thus clearly favors the theory of sacred stone monuments.

Beehive stone niche at Cserépváralja.

However, sitting there above these Szomolya stones, I was not so sure. Perhaps I was uncertain because I had read the writings of Andor Saád, Péter Mihály and Örösi, plus I had seen beekeeping in the rocky cliffs of the aforementioned film, and I did not find it at all unimaginable; the orientation and atmosphere of the whole hive stone complex evoked in my mind's eye a large and buzzing primeval bee. I should add that in the area approaching the hive stones around Szomolya is full of beekeepers.

The Nagy Bábaszék hive stone and the author.

I drove over to Cserépváralja, where I took a detour around the Mango Tető [Mango Roof Hill] and first found the hive stone called Nagy-Bábaszék [Great Midwife Chair]. Its design seemed to me to be absolutely suitable for beekeeping, but at the same time I could definitely feel the beauty and the special power of the place. By this time, something was taking shape inside me, something of which I became certain only at the last stop.

The Nagykúp hive stone and the author's wife.

My last stop was at the Nagykúp [Big Cone] and Kiskúp [Little Cone] in Cserépváralja, where I finally got the feeling that these hive stones could have been both, or alternately over the long centuries: sacral monuments and bee dwellings.

The Kiskúp hive stone

It has happened more than once to people that they forget the function and use of something, and later they spin all sorts of tales about it until the true history is revealed. It also happens that concocted stories become the “real” ones. This mystification may not take away from the original truth; it may even add to it. It is probably also more interesting for most hikers to make a pilgrimage to magical-mystical sacred places than to ancient apiaries. Perhaps I am alone in thinking that these two functions of the stones are not necessarily mutually exclusive. Is it possible that bees were kept here even while these places were so important for spiritual life? Whatever the case, the veil of centuries of oblivion blurs any certainty, but these places are a treasure to be shared and preserved, just as much as our forests and our bees. There is no doubt that, beekeeper or not, these mysterious hive stones are fascinating pilgrimage sites in our country.

Ancient Hungarian Beekeepers, or Bees and the Hungarian Language

What was ancient Hungarian beekeeping like? When we ask this question, a thousand new questions immediately follow. What was life in our ancestral homeland like? Where was the ancestral home? What can we call our ancestral homeland? What language did our ancestors speak centuries before the conquest? How did the conquest take place? But, not to wander too far from the subject of beekeeping, was the knowledge, the culture, the beekeeping (if any), that our ancestors enjoyed in the past, retained in the Carpathian Basin?

These are serious questions, which I have already discussed at length in my book *Hungarian Buddha*, where I showed, citing the outstanding work of Hungarian researchers of Buddhism, how we Hungarians have a very specific picture of the East, of Asia, of the ancestral homeland, and thus of our past, since the time of Sándor Kőrösi Csoma and the researchers who followed in his footsteps, and I could have gone all the way back to Friar Julian.

While I was assembling the research for *The Art of Beekeeping*, and reading the literature more and more deeply, I was suddenly confronted with the ancient stories of the Hungarian people. In my previous article, on the beehive stones, I mentioned the seminal work of József Szabadfalvi, who writes:

> "It is an old practice in the study of Hungarian prehistory to call upon archaeology and linguistics to help us study those periods for which no written sources exist. For the

beekeeping practices of the ancestors of our people, we have no archaeological evidence. The chitinous remains of insects have only recently been investigated in archaeological finds. There are no archaeological remains, and the simple tools of beekeeping could not have survived the millennia. Thus, linguistics is the main source of few, but certain, clues."
(József Szabadfalvi, The past of Hungarian beekeeping, Debrecen, 1992, 8)

Szabadfalvi writes that the originally Finno-Ugric words *méh* [bee], *méz* [honey], *odú* [cavity, hole, nest], *ereszt* [ooze, seep] and *fed* [cover] also exist in Hungarian, and that all the peoples of Finno-Ugric origin use derivatives of these words. He therefore believes that this proves that our ancestors must have been familiar with the most basic concepts and procedures of beekeeping even when they lived together, i.e., about 5 000 years ago.

Using the Historical Etymological Dictionary of the Hungarian Language, he also highlights the following words:

Our word ***méh*** **[bee]** occurs for the first time in 1138 and in 1329 in the form *Mehe*, *Mies*, with the same meaning in our surviving records. In 1219 and 1550 Meh', in 1255 Myhluk. (The second date indicates the year of the surviving copies of the linguistic records.) "Ancient heritage from the Finno-Ugric period: Zyrian *mus*, Cheremis/Mari *meks*, Finnish *mehilainen*. The ancient Finno-Ugric *mekse is related to the corresponding words in the Indo-Iranian languages: Sanskrit maksiká ('bee') and Old Persian mahsi ('fly')."

The word ***méz*** [honey] appeared in 1253 in a linguistic record in the form *Mezam*, in the nominative, and even at that time also meant "honey" In 1280, we find it in the form of *Mezesmal*, but the exact meaning is unknown. Honey is "a thick, sweet food gathered from flowers by bees". Its derivatives are: *mézes* [honey], *mézel* [honeyed], *mézelő* [honeying]. Ancient heritage from the Finno-

Ugric period: Mordvin *med*, Finnish *mesi*; the base form may have been **mete*. It has wide Indo-European linguistic affinities: Sanskrit *mádhu* [honey], Greek *methü* [sweet wine], Russian myod [honey], Swedish mjöd [mead], but it is not possible to decide the direction of borrowing.

Other related words are: ***köpű*** [(butter churning) wooden vessel]; (an older meaning: wicker hive). Derivatives: *köpül, köpülő* [spit, spitting]. Of Old Turkic origin: Tatar *göbe*, Bashkir göbö [wooden chest].

Odú – a hollow in the side of a tree formed by decay; a narrow, dark dwelling, dwelling place. Derivatives: *odvas*, *odvasodik* [burrows, burrowing]. Ancestral heritage from the Finno-Ugric period, but source undetermined. It may be a group of Votyak *udur* [mouth of the beehive], Mordvin *undo* [wooden cavity], Finnish *ontelo* [empty] with the meaning cavity, but another group of Ugric words can also be considered: Vogul *ontar* [womb], Ostyak *onter* [abdominal cavity, chest cavity], where body cavity is the main meaning.

Kas – basket; hive. Word of Slavic origin: Serbo-Croatian, Slovene, Slovak *koš* [basket, beehive].

Viasz – wax, the easily melted material of the honeycomb, used mainly for candle-making. Derivatives: *viaszos* [waxy], *viaszkos* [waxy], *viaszol* [to wax]. The word is of Slavic origin: Serbian-Croatian, Slovenian old *vojsk*. From this, with a vowel that facilitates pronunciation, a more open form and the obscuration of ⇨ i, the Hungarian form evolved, which is used, for example, in the word waxed linen: *vojsk* ⇨ *vojoszk* ⇨ *vajaszk* ⇨ *viaszk*. The colloquial *viasz* was created by the introduction of the consonant shift, adding a *k* in the middle of the word *viaszkgyertya* and similar compounds, but later the sound was dropped.

Raj [swarm] – a great mass of bees, insects, birds; the smallest fighting unit of soldiers: *rajvonalban*; in a swarm line; a group of fighter planes; a unit of children and youths: a higher organization than a guard: a scout swarm, a pioneer swarm. Derivatives: swarming, swarming. Word of Slavic origin: Serbo-Croatian, Slovene, Slovakian *roj* [swarm of bees]; the Hungarian word evolved with a broader use of the word.

And our much later word is ***kaptár*** [beehive] which is a borrowing from the Old German *chafteri* [beehive], derived from the medieval Latin *capisterium* [wooden vessel, cupboard], and can be traced back to the Greek *scaphistérion* [wheat vessel], and even to the root of this word *scaphis* [vessel, bucket]. The Hungarian word ***kaptár*** was formed by phonetic equalization; the *f* ⇨ *p* sound change is due to the fact that before the 14th century, Hungarian only had the f sound at the beginning of a word.

Szabadfalvi concludes from all this that our Hungarian ancestors knew about bees already during the Finno-Ugric coexistence. They knew that they lived in forests, in the burrows of older trees, and that they swarmed in spring and early summer. For a long time, in the vernacular – and I will show later that in old Hungarian bee books as well – there was another Hungarian word for swarming, but today this has fallen out of our language, and we simply call the natural proliferation of bees swarming. Our ancestors knew perfectly well that bees collected honey for themselves in the hives. And now comes the point:

> "Not only did they know these words and their meanings, and thus the concepts, but they used and preserved them continuously for thousands of years. They could only have done this by using and living with it constantly, or at least not with great interruptions."

This means that honey was used to make food and drink, and

wax was also used. According to Szabadfalvi, the words mentioned above also define the technical level of beekeeping, and thus we can say with certainty that the prehistoric Hungarian beekeepers practiced forest beekeeping, so-called honey hunting, which was exploitative. They would seek out wild bees in the woods around their homes, stun them with smoke or suffocate them, and steal their honey. They cut out the sides of the hive and removed the comb. They had to know the bees' annual rhythm, so they could steal their honey at a time when they had already collected a lot, i.e., in the second half of summer or, even more so, in autumn.

This forest beekeeping then developed further, and in time the bees were no longer destroyed, and in fact they were bred, the bee trees were marked (with a carved mark), and only part of the honey was taken. The Slavic and German origins of other beekeeping words such as *kas*, *kosár*, *viasz*, *raj*, *bödön* show our coexistence and interaction with these peoples, and also the changes in beekeeping traditions. For example, in the centuries after the conquest, bees were brought from the forest to the house, and were probably kept in a basket, and much later – from the 18th and 19th centuries – in hives. Thus, beekeeping has constantly changed, developed and evolved.

It is also worth mentioning here that the vocabulary of beekeeping also appears in our ancient place names, and this indeed confirms the historical importance of beekeeping, such as: Méhi, Méhes, Méhtelek, Mézes, Mizdó [honey-giver, Márcadó [mead-maker], Sonkolyos, Kelence [a kind of beekeeper]. These were probably all service villages at one time.

Thus, we can see that, although the Hungarian people knew and worked with bees from the beginning, their culture changed, developed and evolved over the millennia, through migrations and interactions with other peoples, as it continues to do today.

What is particularly interesting is that this forest beekeeping has survived for a very long time, and folklore already describes the traditions in which it is reflected. A fascinating study by Béla Gunda, who also notes that

> "Historical data, geographical names referring to beekeeping, our books on the subject and our ethnographic records prove that this ancient occupation is of great importance for the Hungarian people. The conclusion that Hungarians were already engaged in beekeeping in the Ugric period is not based on a comparison of ethnographic material, but on the etymology of the words 'bee' and 'honey'. Of course, linguistics cannot provide a picture of what beekeeping was like in the Ugric period."
>
> *(Béla Gunda, Research on Hungarian Forager and Predator Agriculture, Institute of Ethnology, Budapest, 1948, 9)*

Beehives from the collection of Béla Gunda (*published by Hungarian Ethnography, Akadémiai Publishers, 2001):*

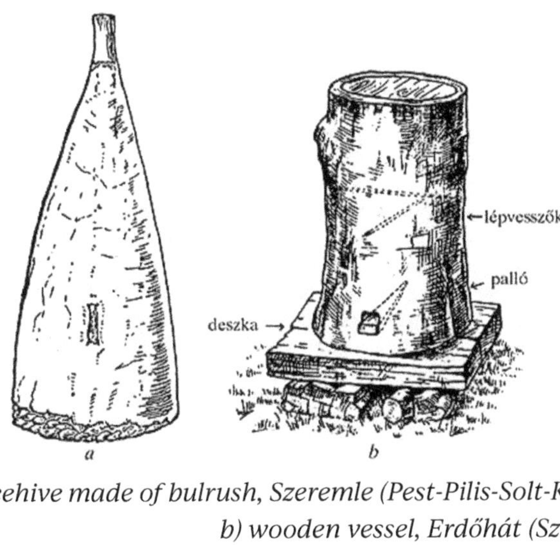

Beehives: a) beehive made of bulrush, Szeremle (Pest-Pilis-Solt-Kiskun County); b) wooden vessel, Erdőhát (Szatmár County.)

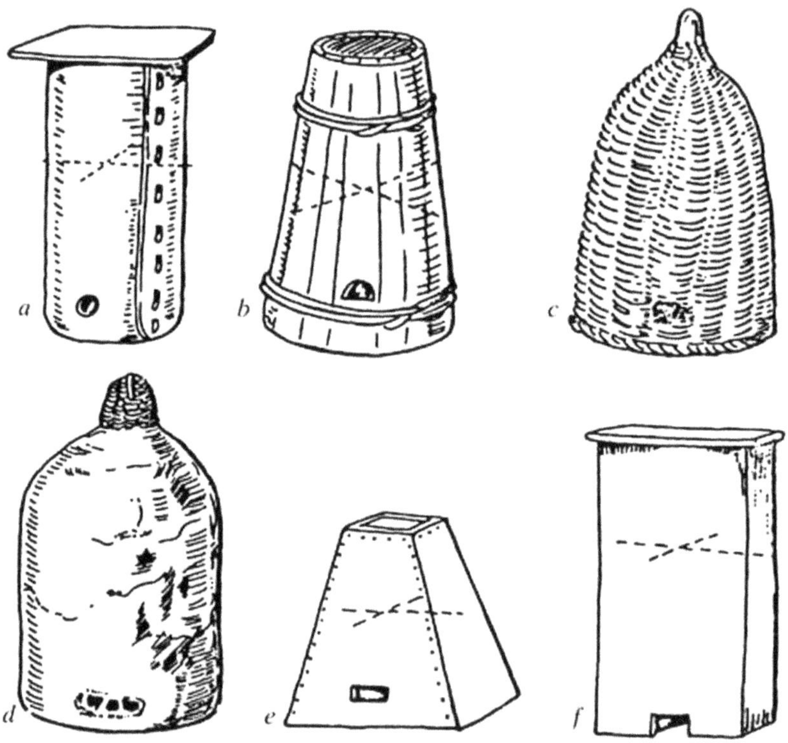

(a)vessel made of birch bark, Magyarigen (Lower Wielkopolska); (b) vessel made of wooden staves, Magyarlápos (Szolnok-Doboka County); (c) vessel made of wicker, Csíkszent-domokos (Csík County); (d) patched wickerwork vessel, Komádi (Bihor County); (e) unframed hive made of boards, Gyimesközéplok (Chisinau County); (f) unframed hive made of boards, Csomakőrös (Háromszék County)

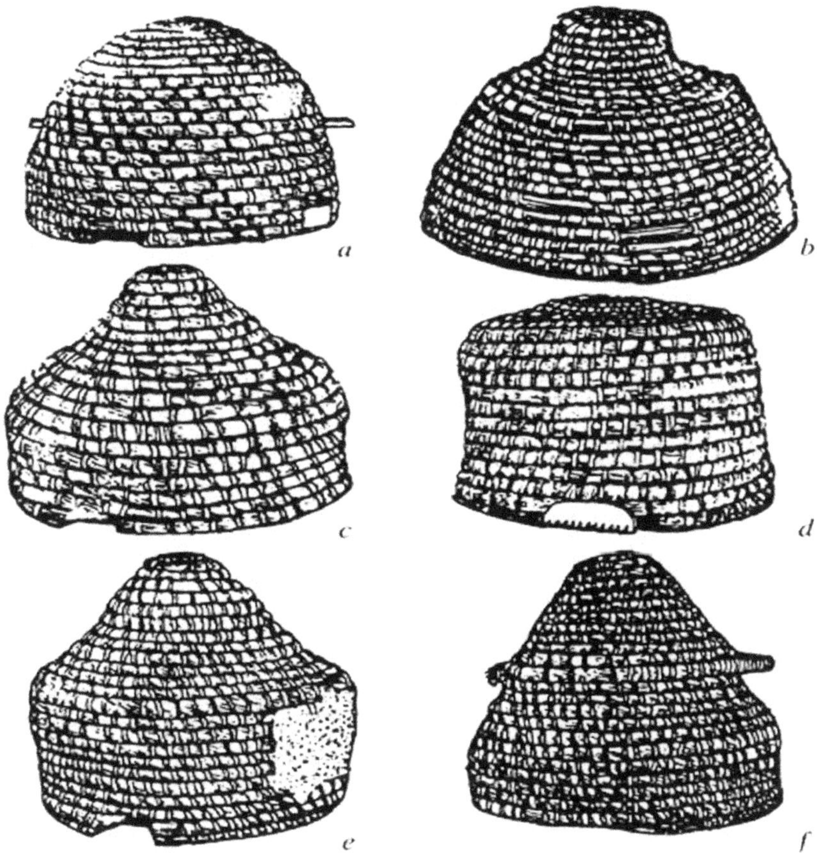

Straw hives from South Transdanubia: a) from Liszó (Somogy County); b) from Pölöskefő (Zala County); c) from Tótszentmárton (Zala County); d) from Letenye (Zala County); e) from Szentmihályhegy (Somogy County); f) from Surd (Somogy County)

In his article Béla Gunda divides the scientific problem of the history of beekeeping into two questions, one about the way of keeping bees, the other about the superstitions and beliefs of beekeepers. The way in which our ancestors kept bees, for example in wooden hives, wicker, straw and bark hives, was not only a matter of natural conditions, but also a reflection of cultural trends. Zsigmond Bátky noticed that different beehives are typical of different regions. For example: "Transylvania has wicker beehives, the northern Hungarian ethnic areas have human-shaped buckets." The Transylvanian people belong to the southern European culture, while the "human-shaped pots" belong to the north-western culture. Such beehives, which are quite astonishing, are also preserved in the Hungarian Ethnographic Museum, and we can find pictures of them in other valuable older beekeeping books, such as those of the esteemed Sándor Ignácz and Baron Béla Ambrózy.

Beehives shaped like human beings, carved stumps from Körmöcbánya in Baron Béla Ambrózy's book.

In Béla Gunda's words:

> "Straw beehives are slowly spreading eastwards from the west, and in the western part of Transylvania they are already beginning to replace wicker beehives, first the typical beehive of southern and then of central Europe. In Transylvania, in the marshes of the Sárrét and in Slavonia, the hives were covered with straw, reed and rush, and are still partly covered [In 1948 – the author]. This custom is a remnant of old European beekeeping. Beehives are kept in wicker containers under the eaves of the house during swarming, at the edges of settlements, in forest clearings."

These early box-like containers are the ancestors of our beehives!

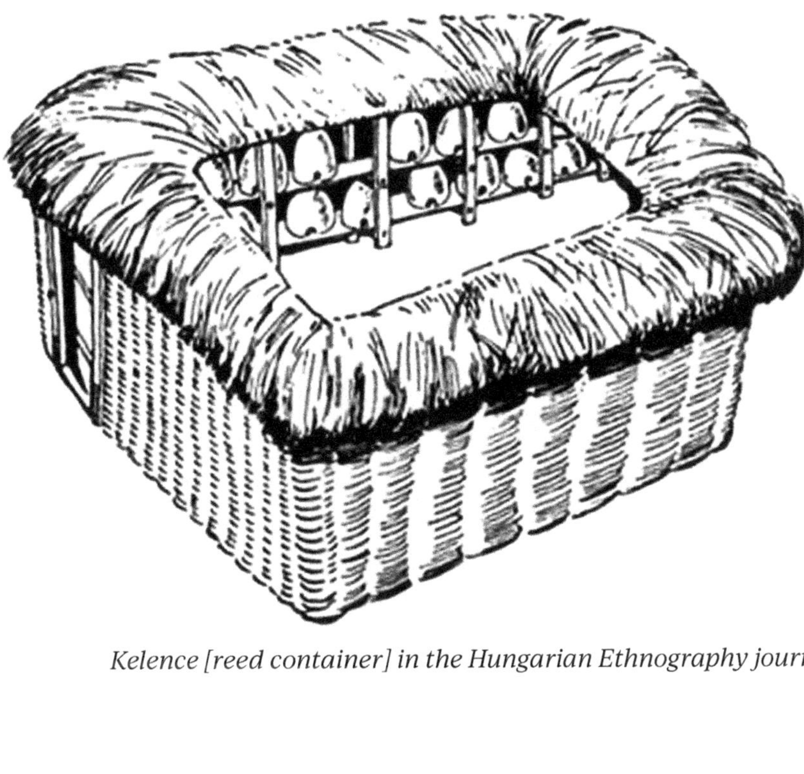

Kelence [reed container] in the Hungarian Ethnography journal.

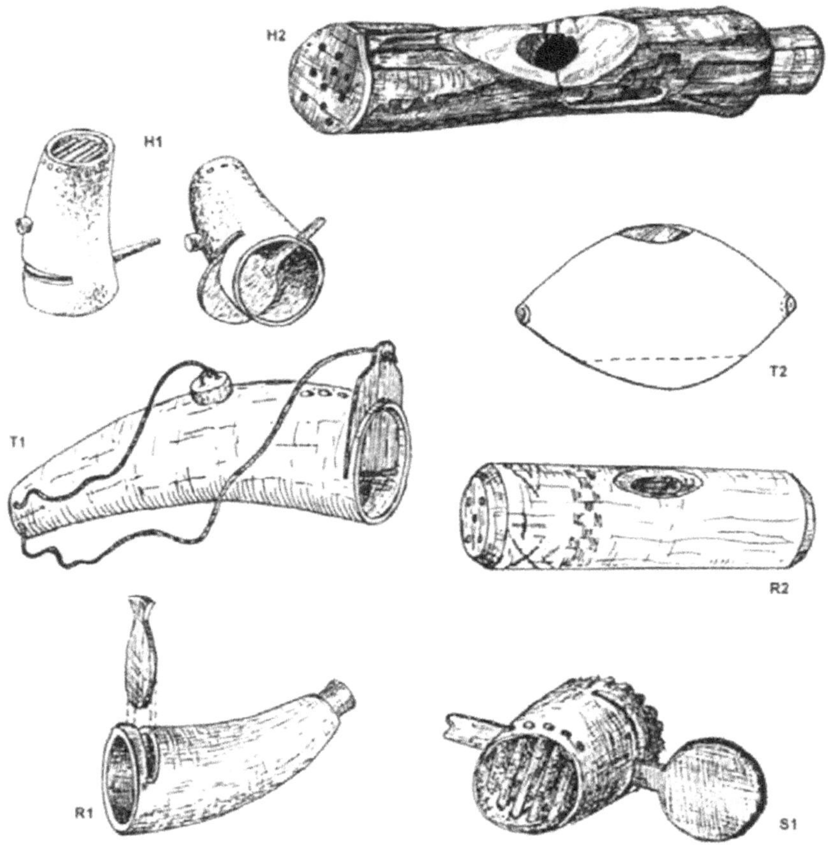

Bee traps from Hungary and Transylvania. H1 Domaháza, Bükk. H2 Pusztafalu, Zemplén. T1 Lővéte, Harghita. T2 Rév, Bihar. R1 Zagra, Radna Hills. R2 Felsőbiharkristyór, Bihar. S1 Hradec, Nitra. Based on the work of Béla Gunda, published by Eva Crane in The World History of Beekeeping and Honey Hunting. New York, 1999. p. 65.

There is a lot more to write about here, but for now I would like to mention only one important and interesting element about ancient Hungarian beekeeping, before we move on to the fascinating world of folk beekeeping. This world is so rich and full of beliefs and magic, for example, how did our forest beekeeping ancestors find bees in the forest? Well, there are interesting speculations, records

Anthropomorphic beehives from the collection of the Museum of Ethnography, with the author.

and even artefacts about this, and this may have been the most extraordinary part of ancient Hungarian beekeeping. Our foraging ancestors would put out a small honey-plate as bait, wait for the first foraging bees to call the others, and then, using a special little device called a *beekeeper's bone* or *honey finder*, capture the bees that flew to them. This was most likely made from ox horn. The bees were released one by one, and by following them, the keepers found the hive. Following the example of Béla Gunda, Eva Crane, the renowned beekeeping historian, devotes a special chapter in her book, which I have already quoted, to the long-established Hungarian folk tradition of forest beekeeping and bee-finding, and its tools, the bee traps.

Later on, the beehive and the stump which contained it was probably taken home. This was how the sophisticated beekeeping of today began, along with its very particular customs, beliefs and quite a rich literature. In my next essay, I'll write about the beekeeping and culture of Hungary in later centuries.

Human-shaped beehives. Wooden figure of a woman in an empire dress, holding a jug in her right hand, with a semicircular flap at her mouth. The color of her dress is red. Notable is the careful elaboration of the hair and headdress, and the naturalistic painting of the face. Height: 160 cm. Inventory number 85.374, Collected by József Ernyey in 1910.

Beehive in the shape of a human: clergyman. Holding in front of himself an open book with two hands. Smooth, blackened hair frames a well-formed face. Height: 155 cm. Inventory number 85.378, Collected by József Ernyey in 1910.

The Apiary,
from Dániel Berzsenyi to Béla Hamvas

Dániel Berzsenyi

1

The legend is quietly spreading, and it has only rarely received adequate attention in Hungarian literature and history. It is a legend we must address if we wish to write about the art of beekeeping. According to this legend, there once lived a Hungarian poet who found himself and his place in the world within the apiary. Not just in language, but in fact. The story goes on to say that he became the most perfect classical poet to have written in this world since Horace, but without committing any particular heroic deeds, without dying a martyr's death on the wrong battlefield, without drinking himself to death in vain. He had a family, four children, and farmed, but mainly he kept bees and wrote poetry.

His poems were never particularly popular, and neither was he. As a result, his memory could not be desecrated on the altar of various delusions; his name was not stamped on flags or printed on money. He remained as he was, a little inaccessible, but somehow full of love for all Hungarians. If someone with a scintilla of intelligence has read this and heard of Dániel Berzsenyi, his essays, his poems, full of classical references, concise and yet distinctly beautiful Hungarian, what may also come to mind is an old portrait of a kind, bearded, warm-eyed old man, with a strangely cheerful sadness.

There are many legends about Berzsenyi, whether we think of his relationship with Kazinczy or Kölcsey, but the story that I find most endearing and interesting is the one about his apiary. Most people know this idyllic image of Berzsenyi from Béla Hamvas's *The Five Geniuses*. Hamvas completely revised this in 1959, having originally written the first version twenty years earlier. More recently, *The Land of the Five Geniuses* was published, based on the original typescript. It would be a particular challenge and of philological interest to compare the changes and differences between these versions of *The Land of the Five Geniuses*, but I will not undertake that now. I want to deal with one legend, the one that Béla Hamvas, already in his early *The Land of the Five Geniuses*, plants, and establishes Berzsenyi's place in the world, in his homeland, and in Hungarian literature. Berzsenyi's place in the apiary. But before we can define Berzsenyi's particular place, we need to illuminate and separate some of the concepts Hamvas touches on. The "place" and the "genius".

What is place?

"The little bee flies to the flower, spreads the petals, sticks its head in and crawls in. What is a flower to a bee? Place. Every place is like a flower, an island, a human body, the earth, a star." – Hamvas writes.

There are places like the whole planet that are in motion; there are places, like a mountain or a valley, that are still; and there are places that are changing like a house that can always be rebuilt, like a room that can always be rearranged. This "place-philosophy" is a cornerstone of Béla Hamvas's theory of genius, because he believes that every place has its "genius".

What is genius?

Perhaps if I said "spirituality", it would be inaccurate, if I said "deity", it would scare many people away. Let's stick to the idea that it's an all-pervading radiance. There are places blessed with stronger genius, holy places and beautiful places, and there are demonic and cursed places. But Hamvas - as many people know well - takes it much further, writing about the "genius" of whole regions and parts of countries. I have already presented and written about the five geniuses, Hamvas's vision of Hungary, and I do not wish to repeat this now, but Hamvas wrote about the *Genius of the South* – I repeat a quotation from the earlier version here:

> "The most intimate place in a mansion is not the library, the dining room, the ancestral hall, the kitchen or the bedroom. The spirit of the South has created something here that no other place on earth has, even in a remotely similar form: the apiary. The apiary is a place of meditation. A summer place, bright, warm. This Southwestern landscape is all asleep: it lures and lulls, like a sleeping siren. Therefore, the South, the symbol of summer, of light: the apiary, the place of meditation, of waking sleep (...)."

Hamvas is not entirely correct in saying that the apiary does not exist elsewhere. The true home of the apiary is Slovenia, but we can find apiaries in several places along the Alps, especially in Switzerland. It is true that Slovenia can be classified as part of the genius of the South, although the influence of the Alps and Triglav is at least as strong and important. Incidentally, one of the most important and beautiful apiaries in Slovenia belonged to Anton Janša, who was invited to the Viennese court by Maria Theresa herself, where he was honored and became the first person to organize beekeeping education in the empire. Anton Janša was an innovator in beekeeping in many respects, but he also expressed himself as an artist, as he was a unique, naive painter who, continuing the special local tradition, painted various images, legends of saints, and stories of beekeepers on the front panels of the hives. I was able to admire his hives, paintings, and stories at the apicultural museum in Radovljica. Here I also learned that he earned the attention of Maria Theresa mainly because of his successful bee breeding (and migration) technology, and that the preservation and spread of the Carniolan honeybee (*Apis mellifera carnica*) as a subspecies, which was later brought to prominence by Slovenian beekeepers.

Anton Janša's apiary with painted beehive front boards in Breznica, Slovenia.

2

All bee species have their good and bad qualities, and in our climate, it seems that the Carniolan bee has been the most popular so far, although climate change may impact this in the long term. The Carniolan bee is also colloquially known as the "grey bee" because of its dark armor, grey hairs on the throat, which are usually short. The color of the queen is hardly different from that of the workers, so it is worth painting it every year to make it easier for the beekeeper to find it among the bustling population. The testes are dark, some are quite black, their bands may appear yellow, but their pattern is dominated by light brown rings. Carniolan bees are considered medium sized, slender, similar to Italian and northern bees. Their characteristics are generally hardy, robust and docile, they overwinter well, are resistant to diseases and adapt very well to different circumstances. Under favorable conditions, they live at least 12% longer than other domestic bee species, and a skilled beekeeper can build up colonies to the main brood. In Hungary there used to be a number of separate types of regional bees, such as the Alföld or Pannonian bee, and the apis mellifica banatica, the latter a cross between the Carniolan and the Italian bee.

The gravestone of Baron Béla Ambrózy at the beekeeping exhibition in Sándorf in 2021. Photo by the author.

It is worth mentioning here that one of the founding fathers of Hungarian beekeeping, Baron Béla Ambrózy, in his book from 1896, *The Bee*, one of the most important books of its time, distinguished between the Carniolan and the Hungarian bee. He describes them very nicely, but I think that by now this has been dismissed in the literature, since the varieties would have been quite mixed in such a dense population of bees. In any case, I was sad to see in the beekeeping museum in Sándorfalva that the legendary Baron Ambrózy's tombstone was lying in tiny fragments on the ground, although at least someone had collected the tombstone's fragments and preserved it for a "grateful" posterity. I have since read that his tombstone was put back together when the remains of the baron and his family were reburied, but the beekeeping museum in Sándorfalvi was emptied at the same time. Wherever it has gone, this tombstone is a sad symbol of pre-war Hungarian beekeeping, such as Berzsenyi's beekeeper practiced, because he most certainly had bees of this kind in his apiary.

Hamvas thus continues the story and creates a legend:

"But the apiary, besides being a place of waking sleep, is more than that: it is poetry and philosophy and morality and behavior and metaphysics. And Berzsenyi is the philosopher, the moralist, the poet of the bee. You have to imagine him, after lunch on a July day, in the scorching heat and the sparkling sun, slowly crossing the courtyard with Horace under his arm and opening the door of the apiary, fifty paces from the house. A small room. He enters. The bees buzz around the hives, flying in and out of the window, in and out, forever in a circle. Apart from the hives, there are only pine chairs and a table, nothing else. He sits down, looks out over the blazing garden, then turns to the bees and forgets to open the book. Suddenly the world of external images disappears, and one sinks into a place from which one can no longer see garden or hive. He sits motionless and, something inside him begins to expand all that he has lived, noiselessly, like dough, mixing and spitting out experiences: loves, childhood, parents, houses, thoughts. Something is at work in him, turning his life around, - but in what unexpected colors; how deeply he sees now all that he saw then so briefly and foolishly. Now it is scarcely a human life, when in his ears only the buzzing of little insects hum in the distance, and how this humming is like summer and heat and light and slow time flowing oily, rising in him and in his dreams and the future and the beautiful and the great, ambition, triumph, kneading and washing like cloth and pouring thoughts from one minute to the next. Thus, he devours his life in this Zeusian solitude. . This is meditation. (...)"

(Hamvas Béla, The Land of the Five Geniuses, (1940) Medio, 2021, 29.)

What could have been Béla Hamvas's source? Did he visit Nikla? Could he have seen Berzsenyi's apiary? Who first mentioned Berzsenyi's apiary? What happened to the apiary?

3

Béla Hamvas may have visited Dániel Berzsenyi's village of Nikla, but it is impossible to prove this; it is more likely that he read about it. Researching items in the press, I found a wide variety of publications even to this day about Berzsenyi and bees. For example, they speak of swarming bees settling on the poet's hat, but he didn't notice because he was writing a poem. Or they say that the inhabitants of Nikla knew that when smoke from a pipe rises over a beehive, that great thoughts are born. I have even read that there were so many beekeepers around Nikla because Berzsenyi himself encouraged the locals to take up this noble activity, so the knowledge and love of bees was passed down from father to son, through generations. Otherwise, this isn't particularly unique, and it is indeed rare that someone without a family tradition just starts beekeeping. In such cases a good mentor is very necessary, especially in the beginning. Beekeeping is not something you can learn from books and instructional videos. You need a mentor with experience.

Returning to Berzsenyi's apiary, I must first of all highlight two useful resources. First, I quote from Kálmán Smidéliusz's essay on Dániel Berzsenyi's home:

> (on Dániel Berzsenyi's Apiary) "the 42x12-meter apiary, with an external wall, 65 cm thick, has withstood the elemental forces of time for a long while, but was severely damaged during World War II. Going around the house, one reaches the edge of the garden. Here once stood the apiary that Béla Hamvas called the most intimate space of the noble mansion, the 'place of wakeful sleep' in the southwest. Berzsenyi often came here from his house, which was always cool, dark and smelled of smoke. Right here, behind the house, because only the bourgeoisie would sit outside the gate to watch the street and smoke a pipe. Here he could be far enough away from the clamor of his four children and the chatter of his wife, but still close enough to be able to give orders and take action. Here, in the apiary, he could lose himself in a book or find himself in a poem or a letter. And if he reached for his wine glass, he could look up and gaze with satisfaction around every corner of his estate with the eyes of an owner. No one has sat out here for a long time. The present visitor sees an empty, overgrown nothing. And that only a few stray wasps from the nearby orchard visit the site of the once-vacant apiary (...)"
>
> *(Kálmán Smidéliusz, Dániel Berzsenyi's Homes. Irodalomismeret, 2011/3, 86.)*

There is a lot of valuable information to be gleaned from this little piece of information, especially if we add the short but valuable article by László Kenéz in which he writes:

> "Who would not decide, after reading the lines (by Béla Hamvas quoted above), to go to Nikla, to the Berzsenyi mansion, and smell the air in the house and the garden, but above all to see the apiary? But there is a glimmer of light,

because, strangely enough, at the very end of the designated walk, the apiary appears. Sadly, there is only one photo of it, inside in the corridor of the mansion. Here it is."
Dániel Berzsenyi's apiary.

The only picture we have of Dániel Berzsenyi's former apiary in Nikla.

Kenéz continues:

"Let's go out again, let's at least find where the apiary stood. Finally, we have the location from the old picture. And then, in this very place, everything becomes clear as day. The apiary was the link between the grove and the mansion. Now that it is gone, the lens through which the spirit of nature could be projected into the rooms of the house are missing. Hence the sharp contrast between the glorious garden and the barren interiors. The apiary was an oculus, a magic eye. It is enough, then, to stand in the cool place of the apiary to catch a glimpse of that which is really there, instead of the neon-lit, absent Berzsenyi."

(László Kenéz, The Apiary's Cold Spot (A méhes hűlt helye). Kultúrpart, 8 August 2008.)

The apiary disappeared after the war; it had been destroyed. It must have been unique in that Berzsenyi probably really did take a nap in it after lunch, much as Baron Béla Ambrózy wrote in his book:

> "Apiaries are used to house bees permanently. The apiary is either open or closed. The apiary has a place not only for the hives, but also for the various beekeeping tools. In fact, the apiary is often the beekeeper's resting place, when it also contains a couch or bed."
>
> *(Baron Béla Ambrózy, The Bee. (1896) Fourth edition, 1992, 280.)*

It's worth looking at the apiaries in Baron Ambrózy's book; they were exciting and imaginative buildings, and particularly beautiful. These apiaries were certainly different from the beautifully painted apiaries of Slovenia, different from the ones that were developed during the 20th century for travelling beekeeping and, above all, different from the beekeeping containers which were developed practically for production. But they foreshadowed the apiary houses that are becoming increasingly fashionable today, and of which more and more are being built in Germany, Slovenia, the Czech Republic and Transylvania. You can actually sleep in them without being disturbed by bees, be refreshed by the sweet fragrance and spend long periods of time near the bees, enjoying their buzzing, just as Dániel Berzsenyi did two hundred years ago. There are many explanations why this is healthy and good for you. I will not go into that now, but rather return to the question of where and from whom did the legend of Dániel Berzsenyi's apiary originate?

4

If we look carefully in the contemporary press, we find, after much research, that the earliest mention of this apiary is on June 14, 1860, when Berzsenyi's memorial obelisk, under which his ashes rest, was unveiled with great ceremony in the village. The event was very important for the Hungarians of the time, as a memorial column was erected to a great Hungarian poet after a lost revolution and war of independence, but before the Reunification, in the Bach era.

As they say:

"Thousands of people flooded the country roads; the cemetery in Nikla could not accommodate the crowds. But they, both those in the cemetery and those outside, sang the National Anthem and the 'Appeal' [the 'second national anthem' with lyrics by Mihály Vörösmarty] loud and clear.

"It was a time of autocracy, the Bach era - an obelisk thrusting into the sky on the small hilltop of the Nikla cemetery represented the nation.

"The newly erected tomb of Dániel Berzsenyi bears the inscription: '...the brave shall not fear...' Four lines from Berzsenyi's great poem:

"'The brave shall not fear the moss of time,

They shall break from their coffins; they shall call to heaven!

(...)'

"According to the report of the ceremony, at the monument's unveiling, a bearded, elegant man in traditional ceremonial dress suddenly appeared. With a sword at his side, he stepped up to the obelisk and, speaking of Dániel Berzsenyi's patriotism, raised his voice. At first, the people just murmured and whispered, until - as if a bomb had exploded – the Kossuth song burst out of them:

> “’Lajos Kossuth has sent word,
>
> His regiment has run out . . .’
>
> “In the orators, they recognized Lajos Kossuth, who was returning home from exile. Who as suddenly he had come, suddenly disappeared. In time, fortunately, because the emperor’s men had already sent for the mercenaries...”
> *(Excerpt from the commemorative speech of János Dénes, delivered on the occasion of the 184th anniversary of Berzsenyi’s death in the Nikla cemetery, 22 February 2020.)*

We don’t know whether Lajos Kossuth actually appeared in person from exile at the time, but I have read up on who wrote about this great event and what they wrote. I must admit that my heart was filled with joy and unexpected surprise when I read two long reports about this in the June 1860 issue of the *Pesti Napló*. In one of them I read this:

> “In the background of the garden stands an abandoned apiary, and this was Berzsenyi’s sanctuary, where he retired from noise to poetry, and his poems were mostly written amidst the buzzing of bees and the whispering of the wind, even with thunderstorms and lightning.”
> *(June 14 (anniversary of Berzsenyi’s death). Pesti Napló. 1860/6.)*

And below the article, the author’s name is Zsigmond Kemény, that remote and inaccessible pinnacle of Hungarian prose literature. Zsigmond Kemény, whose abandoned and dilapidated castle in Pusztakamarás [Cămărașu, in Romania] I visited, and whose first, forgotten novel I wrote about.

From the book "The Bee" by Baron Béla Ambrózy, the model beekeeper of the Pilis Region Beekeeping Association in Budakalász.

A wooden pavilion-like beehive on a brick base, from Béla Ambrózy's book The Bee.

From the book "The Bee" by Baron Béla Ambrózy, the model beekeeper of the Pilis Region Beekeeping Association in Budakalász.

In the book of Baron Ambrózy, the beehive.

Hexagonal, pavilion-like beehive for 45 bee colonies
From Baron Béla Ambrózy's book The Bee.

Pavilion for 100 bee colonies
From the book The Bee by Baron Béla Ambrózy.

So that's how stories come full circle, I thought. I don't know if Béla Hamvas read Zsigmond Kemény's article, but I do know that Hamvas loved and respected Kemény; at least in his few writings on Hungarian subjects, he certainly speaks positively of him. In any case, this is the first source of such a mention of the poet, poetry and the apiary together, and probably the first in Hungarian literature.

Zsigmond Kemény saw Dániel Berzsenyi's apiary, Béla Hamvas could have seen it, as it was only dismantled after World War II, but at least he knew about it and understood its significance. And we can marvel that two hundred years ago there was still a building in the country where beekeeping and poetry met. There was a house that was a "magic eye" through which one could peer into the world and into the homeland, that was both ancient and ahead of its time, and whose genius slumbers here, surrounded by the buzzing of bees, somewhere in the golden dust of the centuries, waiting to be rebuilt, now with the knowledge of the new century but with respect for the old traditions.

Béla Hamvas in his garden in Szentendre in the 1960s. Photo by Aladár Záhonyi.

Bees and the Wizard Beekeeper, or Why "Beekeeping is the Poetry of Agriculture"

In the middle of summer, my neighbors see nothing but my atelier filled with honey, and me walking around in a white hat and white clothes, surrounded by fragrant wisps of smoke, looking around and tidying up my apiary. Luckily, I live in a good neighborhood, among good people who support me in my beekeeping. They also get more fruit and vegetables as a result, and the locals also know to value local honey. Still, I feel that even before, as I am a poet, they looked at me with amazement and perhaps even whispered, "What a strange creature this Tibor is!", but now this amazement is accompanied by a kind of reverence. They often say things like, "Oh, you're a beekeeper?" and "How are the bees?" and when they do, I always tell them stories, and they listen with wide eyes. The other day, one of them asked me, "Don't they bite you?" I tell them that bees don't *bite*, they *sting*, and that this year I've been stung 98 times, which seems a bit much compared to last year's 43, but I'm almost looking forward to it, and it doesn't swell up at all, and I hope I'll reach 100 by the end of the season. Of course, when they ask me many times, I usually say, "No, they don't sting." And they say, "Oh! Wow!" And I seem like a wizard in their eyes. Wizards have big white coats, hats, smoke surrounds them, they wander around dangerous places, just like Gandalf in Tolkien's novels. But I'm more like a hobbit from the county who loves peace, cheerfulness, beauty, and of course honey, than a great wizard who gets into one adventure after another. And here we come to a very important and very delicate part of the art of beekeeping, namely that the beekeeper is also a bit of a wizard, and this too goes back to the distant past. Before you dismiss this idea, let's take a look at the history of beekeeping, especially in Hungary, and then we will understand why this image of the "wizard beekeeper" or, as folklore records, "the devilish beekeeper" came about. All this is closely related to the fact that for many centuries, consciousness and instinct did not separate in beekeeping, which is perhaps why beekeeping is such a wonderful interplay of consciousness and instinct.

Ian McKellen, who also plays Gandalf, stars as an elderly and retired Sherlock who takes up beekeeping in the film Mr. Holmes.

"Beekeeping is the poetry of agriculture..."

wrote Béla Gunda, the famous ethnographer whose works are particularly valuable to us beekeepers, as are the books of József

Szabadfalvi and Eva Crane.

All three of these authors studied the history of beekeeping, especially its ethnography, in great depth and detail. I have already quoted all three in this series of essays since all of them also studied beekeeping books and beekeeping customs. They also emphasized that old reference books on the subject provide a wealth of information about beekeeping customs of the time, especially superstitions. This is the point at which we beekeepers can learn a great deal from the beekeepers of past centuries, and indeed about ourselves. For example, when delving into old beekeeping books and smiling at a strange custom or superstition, we should ask ourselves whether these customs might have been based on valuable natural observation, and whether we, too, have similar customs.

Over the centuries, instinctive forest beekeeping and honey hunting were slowly replaced by increasingly conscious beekeeping, namely by cutting down marked bee trees and taking them home. This allowed our beekeeping ancestors to observe the life of bees much more closely. I would like to note here that Aristotle built a glass-walled beehive in ancient times so that he could better observe the life of bees, but the real breakthrough in the development of beekeeping came with the introduction of removable frame hives. Until then, beekeeping was mainly carried out in tree trunks, rock crevices, logs, hollows, and hives, which made it difficult to inspect the bee colonies week after week and observe the exact course of their lives. Béla Gunda writes at length about this process in his excellent *study Beekeeping among the Hungarians (Méhészkedés a magyarságnál, Agria yearbook, 1991-1992, pp. 303-368),* and in two other writings he also highlights the transition from gathering honey to conscious beekeeping. I could write at length about Gunda's wonderful observations, as he was one of those ethnographers who gained their knowledge both in the field and in the library. Suffice it to say that he, like Szabadfalvi, mentions that at least fifty books and studies on beekeeping were published in Hungarian from the eighteenth century onwards. These books vividly illustrate how beekeeping became recognized as a craft.

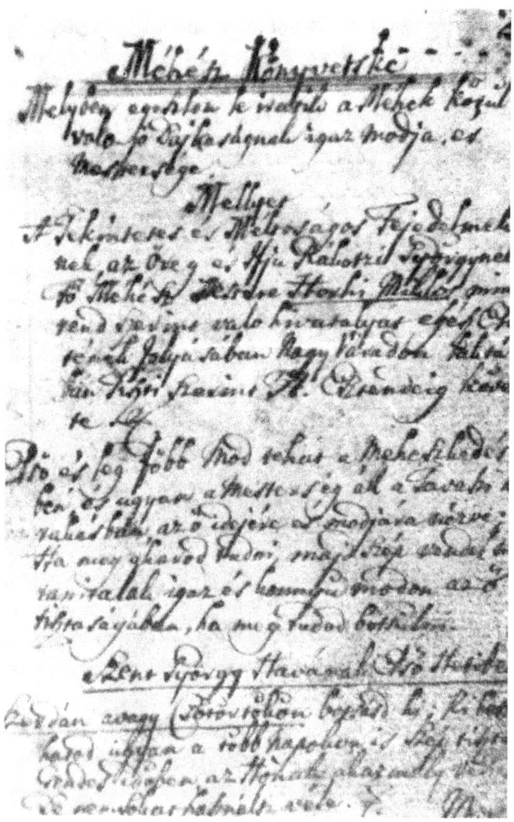

Title page of a copy of Miklós Horhi's handwritten beekeeping book from 1720.

The oldest Hungarian-language book on beekeeping was published in 1645 in Nagyvárad (Oradea) by Miklós Horhi, chief beekeeper to György Rákóczi. Only handwritten copies of this book have survived. In addition, the following printed works are also noteworthy: Lőrinc Pálfi *(Erdélyi méhecske [Bees of Transylvania], Kolozsvár, 1762)*, Gyula Szigeti *(Méhész könyv [Beekeeping Book], Nagyenyed, 1763)*, Márton Széki Vesmás (*Erdéllyi méhes kert [Transylvanian Bee Garden], Kolozsvár, 1774)*, György Handerla *(Új méhész [New Beekeeper], Pozsony, 1775)*, András Ravazdy *(Méhtolmács [Bee Interpreter], Diószeg, 1791)*, and Antal Veszelszki *(A magyarországi méhtartás {Beekeeping in Hungary], Vác, 1795)*. Béla Gunda mentions that

> "These authors (mostly rural farmers, clergymen, teachers) write about their experiences among peasants and also share a wealth of folklore. Some of the works reflect the advice and ideas of classical authors (Virgil, Columella) and contemporary foreign beekeeping literature."

John Gedde's work *The English Apiary* (1721) was a very influential book in Europe at the time and was published in three editions in Hungarian translation under the title *Angliai Méheskert* (1759, 1768, 1781). It was revised by István Vannai in 1779 to suit Hungarian conditions. György Csáti Szabó's book *Magyarországi méhesgazda (Beekeeping in Hungary, Debrecen, 1792)* was a major step forward in Hungarian beekeeping and was followed by a series of books that increasingly sought to separate conscious beekeeping from superstitions. that most contemporary beekeepers instinctively followed, unaware of whether they were useful or not. I was able to look through several of these books at the National Széchényi Library and the Somogyi Library in Szeged.

It is worth noting that during the reign of Maria Theresa (1740-1780), beekeeping was developed through official measures and professional guidance. In 1770, a beekeeping school was established in Vienna under the leadership of Anton Janša, whom I mentioned in my previous essay (entitled *The Apiary*), and Hungarian students were also recruited. In the same year (!), a beekeeping inspector was working in Hungary, giving lectures on bee management and encouraging counties to adopt modern beekeeping practices, as Béla Gunda also emphasizes. It is worth mentioning that a beekeeping school was founded in Szarvas in 1771 (!), where teaching was free and where lessons were still being held in 1805. All this is linked to the name of Sámuel Tessedik, as is the knowledge of how to keep bees without suffocating them. In Keszthely, beekeeping was included in the school curriculum in 1797.

Old Hungarian beekeeping books in the Somogyi Library in Szeged.

One of my particular favorites is the former parish priest of Szany, János Nagy, whose work *Méhi gazdaság (Beekeeping)* was published in Győr in 1786. In it, the author summarizes the science of beekeeping in Hungarian verses adapted from Ovid and even adds moral advice and guidance. For example, he writes about theft:

"Here it is necessary to first recognize the thief, who comes with anger to the hive. He comes in large swarms and early.

He is blackish in color, because the bees that feed only on honey are browner around the waist and belly.

He knows the hive and enters it with great force.

They pull each other and kill each other with great anger.

Seeing this, take some mud, which is always mixed with cattle dung and ash and mix it lightly.

Make the hole smaller, just big enough for a bee to get in, and you will give it strength."

And if the robbery breaks out, when the stronger bee colonies begin to plunder the weak, we cannot advise much else, although our tools are different, and perhaps even our solutions, and there are those who move the hive away, but first we narrow the exits to make it easier to defend, and secondly we try to mask the smells that trigger thc robbery with some strong-smelling substance and smoke. This is why we need to feed them in a way that masks odors, and why we need to strive to have families of equal strength, as well-fed as possible, and without any motherless bees. Silent theft can still occur, and this is where a very interesting folk belief comes into play.

Namely, that the wizard beekeeper, or the devilish beekeeper, used his own bees to steal other beekeepers' bees, and even lure away their swarms, because, as they said, "a queen bee's stinger was stitched into his finger." The aforementioned beekeeping books contain lengthy passages on how to protect our bees from others, especially from other devilish beekeepers. For example, to make our bees "brave" and "biting," they must be released into a wolf's den. To protect our hives from damage and the evil eye (especially from envious individuals), we must display horse skulls. Béla Gunda traces the latter custom back to the Hungarian conquerors. The most astonishing customs relate to how to deal with bee thieves, as this has unfortunately been a problem for many centuries. According to one such custom, a beekeeper whose bees have been stolen must fast, and while he fasts, the thief will fall ill, and if he does not return the bees, he will die. There are numerous stories about bee thieves, for example, that in the past they were punished by having a basket full of bees poured over their heads.

The earliest beekeeping book also mentions a custom related to the beekeeper's urine, which goes as follows:

> "So that the bees do not sting you, sprinkle them with your urine, and then smoke yourself with pine resin so that they will not sting you,"
>
> and
>
> "When you want to knock them down, spray the hive with urine, then smoke it with pine resin, and rub it with its hard flowers or green leaves..." –

writes Miklós Horhi in 1645. Elsewhere we read that the beekeeper should urinate into the bees' drinking water, or urinate on the side of the beehive, and so on. At first, I smiled at this, but then I thought about it, and I'm not sure that it's a completely unfounded superstition. After all, I have observed countless times that my bees prefer to land on stagnant and living water rather than clean water. There are serious studies showing that bees collect something from animal excrement, and Dr. Hegedűs gave a very important and interesting lecture not long ago on why chlorinated, clean tap water is not good for the intestinal flora of bees. I had no choice but to try it out in the spring. I carried water from a mountain spring half an hour's walk away for my bees and placed it next to clean tap water and spring water. I even urinated into a small bucket of rainwater. I also have a small garden pond, and then I watched to see which of the four water sources my bees preferred. You can guess.

The following belongs to this topic, too, but the death of a beekeeper and the wide range of beliefs and superstitions associated with it should be the subject of a separate article. We could write at length about the many traditions and beliefs surrounding the wizard beekeeper or devilish beekeeper in Hungarian folklore.

Many beekeepers were considered wizards, and even in specialist books it took centuries for instinctive, superstitious beekeeping to be separated from conscious practice, while beekeeping became an increasingly difficult craft, as Béla Gunda writes:

> "Beekeeping – which is said to be the poetry of agriculture – began to decline in the second half of the 19th century because honey was increasingly replaced by sugar. Wax was no longer used in candle making, and gingerbread baking also lost its importance. (...) With intensive farming, fields and meadows are plowed, stubble is cleared, forests are cut down, and chemicals are used recklessly, reducing the area of good bee pastures. Recently, however, mainly through the efforts of beekeepers themselves, great care is once again being taken to preserve beekeeping."

Bees and Death, or What to *Tell* the Bees

1

The most unusual belief and custom related to bees is what is known as "telling the bees", a tradition which has survived primarily in Slavic, German, and English-speaking regions. At the end of my previous essay, I mentioned that when a beekeeper died, one of the family members would go out to the hive and "tell the bees" that their master had passed away. Since my previous two articles dealt mainly with Hungarian customs and beliefs, it is worth touching on this very unique and well-known tradition, as it has inspired numerous poems, paintings, stories, and even me on one occasion, making it indispensable in *The Art of Beekeeping*. If we search the internet, we find that "telling the bees" of the death of a beekeeper was probably an ancient Celtic custom that was passed down for thousands of years, but this is only true insofar as humans have associated many mystical and religious ideas with "the Daughters of the Sun" since the beginning of time.

Hilda Ransome, one of the most authoritative researchers on beliefs related to bees, writes that

> " . . . the custom of telling the bees of a death originated in the idea that the bees were 'souls,' or that they were creatures who could fly up into the heavens from whence they had come. The custom is very widely spread; we find it all over Central Europe, in the British Isles, in America, where the settlers from Europe carried it, and the first allusion to it can be traced to classical times; for in the *Greek Anthology* we find

the bees are to be told of a death, though here the purport of the 'telling' is not the same as that attributed to it later . . . "
(Hilda M. Ransome, The Sacred Bee in Ancient Times and Folklore. London: George Allen & Unwin, 1937.)

This idea is qualified by Eva Crane, who believes that this peculiar custom is related to "the secularization of bees," that is, that after the Middle Ages, beehives appeared in more and more ordinary households in addition to and sometimes instead of in church estates. While in medieval Europe, including here in Hungary, it was mostly monks who kept the bees around the abbeys, so the Church controlled beekeeping, and Christian ideas and beliefs became linked to it. But after the Middle Ages, with the decline of the church's influence, and especially with the advent of colonial sugar cultivation, a lot changed. For example, sugar from the Caribbean appeared in large quantities and at much lower prices than honey on the European market, and beeswax candles, which had previously been used mainly in church ceremonies, became less important. Small apiaries, mainly owned by teachers, priests, small landowners, and minor nobles, appeared, usually close to family homes and in gardens, since beekeeping typically did not require a large area. So, there were mainly economic and historical reasons for beekeeping passing from the church into private hands, and one of the main reasons was the arrival of sugar. The large quantities of cheap sugar not only transformed the history of beekeeping, but also our world. We could analyze at length how colonization and the suffering, exploitation, wars, and slavery that followed are closely linked to the increasing consumption of sugar and, of course, the decline of honey. For now, however, let us content ourselves with noting that, although it is possible that the ancient Celtic peoples had some beliefs about bees, most serious researchers who have studied bees in depth agree that the first records of the custom of "telling bees" date from around the 16th century. Our first contemporary source of the time, the neo-Latin author Joachim Camerarius (1500-1574), wrote:

> "Who would believe, without superstition, that most bees die in their own hives when the beekeeper dies? (...) And yet I myself have seen this happen several times, even to people who were not superstitious in the least."

There are numerous records from 17th-century France stating that when a beekeeper died and the family mourned, the hives were covered with black velvet, otherwise the bees would also die after the beekeeper. This custom was particularly widespread in the nineteenth century, primarily in the United States, where the best-known trace of this strange tradition is the poem *Telling the Bees*, written in 1858 by the American poet John Greenleaf Whittier (1807-1892).

Whittier is deeply loved in Hungary because, after the failure of the 1848-49 Revolution and the War of Independence in 1851, Lajos Kossuth went to America. There, it was Whittier who welcomed him with his poem *Kossuth*, which can be found in Miklós Radnóti's excellent translation. In this poem, the poet demanded the liberation of slaves and referred specifically to the emancipation of serfs in Hungary; later, it was precisely this demand that caused the Southern states' secession and the American Civil War from 1861 to 1865. In addition to farming, beekeeping, and writing poetry, Whittier was also a political activist who did a great deal for the abolition of slavery. He never owned slaves and was an elector for Abraham Lincoln in 1860 and 1864. He worked as an editor and critic, and his best-known book is *Snow-Bound*, published in 1866, in which he expressed the sincere desire of people for an idealized rural life in the aftermath of the bloody Civil War. (I was the first to translate Whittier's poem into Hungarian, and I heavily reworked it, infusing the translation with my own voice, feelings, and memories, making in effect, the poem my own, but here is Whittier's original):

Telling the Bees

Here is the place; right over the hill
Runs the path I took;
You can see the gap in the old wall still,
And the stepping-stones in the shallow brook.

There is the house, with the gate red-barred,
And the poplars tall;
And the barn's brown length, and the cattle-yard,
And the white horns tossing above the wall.

There are the beehives ranged in the sun;
And down by the brink
Of the brook are her poor flowers, weed-o'errun,
Pansy and daffodil, rose and pink.

A year has gone, as the tortoise goes,
Heavy and slow;
And the same rose blows, and the same sun glows,
And the same brook sings of a year ago.

There's the same sweet clover-smell in the breeze;
And the June sun warm
Tangles his wings of fire in the trees,
Setting, as then, over Fernside farm.

I mind me how with a lover's care
From my Sunday coat
I brushed off the burrs, and smoothed my hair,
And cooled at the brookside my brow and throat.

Since we parted, a month had passed,–
To love, a year;
Down through the beeches I looked at last
On the little red gate and the well-sweep near.

I can see it all now,–the slantwise rain
Of light through the leaves,
The sundown's blaze on her window-pane,
The bloom of her roses under the eaves.

Just the same as a month before,–
The house and the trees,
The barn's brown gable, the vine by the door,–
Nothing changed but the hives of bees.

Before them, under the garden wall,
Forward and back,
Went drearily singing the chore-girl small,
Draping each hive with a shred of black.

Trembling, I listened: the summer sun
Had the chill of snow;
For I knew she was telling the bees of one
Gone on the journey we all must go!

Then I said to myself, "My Mary weeps
For the dead to-day:
Haply her blind old grandsire sleeps
The fret and the pain of his age away."

But her dog whined low; on the doorway sill,
With his cane to his chin,
The old man sat; and the chore-girl still
Sung to the bees stealing out and in.

And the song she was singing ever since
In my ear sounds on:—
"Stay at home, pretty bees, fly not hence!
Mistress Mary is dead and gone!"

John Greenleaf Whittier

2

It is particularly interesting that, according to records, people not only talked to bees, but also sang and danced to them. It was common practice not only when a beekeeper died, but also when someone got married or was born, i.e., when the family received good news, to go out to the bees and tell them. Even in our country, there are records of people giving bees some of the wedding cake so that they could rejoice with the family and, of course, produce more honey.

It is worth asking why people started talking to bees. What could have been behind this strange custom? It is a fact that when a beekeeper died, the bees would indeed move away and die, either because the family could not care for them as well as their former owner, or because there was something to this belief and a close bond had formed between the beekeeper and his bees. I will leave it up to my readers to decide; I would not rule out either possibility. This is especially true in light of the rather mystical story recorded in 1956 upon the death of an American beekeeper of Polish origin. John Zepka was a respected old beekeeper in the Berkshire Hills, and when he died, a swarm of bees appeared at his funeral and circled above his grave. Several newspapers reported on the incident, including the *New York Times*, which described it as a strange story "unlike anything we have ever seen before."

All this may indeed be related to what Eva Crane noticed, namely that bees kept by families became part of family life over time and were regarded as pets, or, as we say in Hungarian, as "livestock", i.e., animals closely related to the family. A dutiful farmer feels responsible for his animals, loves them, and cares for them, because he himself depends on them. Of course, dogs and bees hold different places among domestic animals, but I find it very interesting that there are many stories about dogs rejoicing with their owners and mourning them when they die. Do bees recognize their beekeepers? What kind of bond can develop between bees and beekeepers?

I should note that when the old beekeeper, my friend's father, from whom we inherited most of our first bees, died in the fall of 2020, before I could bring his bee colonies to my place, I went out to them myself and recited my poem "The Old Beekeeper's Smoker" to them, saying, "Your father is dead, but don't cry, / I will be your father and continue / where he left off."

Bees "mourn" when the central figure of the colony, the queen bee, dies, and as is well known among beekeepers, they make a strange, unmistakable sound, which the literature describes as "the bees crying." They can also be happy, for example, an attentive observer can see the "joy flight" in the golden sunshine at the end of a beautiful day after feeding. Unfortunately, this is also noticed by other bee colonies, and robbery can easily occur, which is why it is better to feed them at dusk.

I don't know if bees "understand" what we say to them, but we can say for sure that if we are impatient and angry, they are more likely to sting. However, I have found several reports that it has happened more than once that half or even a whole colony has died after the death of the beekeeper. In fact, a very experienced beekeeper friend of mine told me that when he was emotionally disturbed and unable to concentrate on his work with the bees, he lost most of his colony and had to start all over again. Our bees seem to like it when we talk to them, even if some people think we are crazy. They probably like poetry too, because they all survived the first winter beautifully, perhaps because I told them what they needed to know, or perhaps because their former owner had fed them well, and they were free of mites and nosema, at least as far as possible. And although I suffered losses in the early spring of our second winter, just as the Russian-Ukrainian war broke out, and there were low points and grief, we still got back on our feet, grew stronger, had a bountiful honey harvest, and most importantly for me, I learned from my mistakes. And now, as I write this, I am wintering with twenty-three bee colonies, which, despite the severe drought, have done their indispensable work of pollination so beautifully.

After all, although we love bees because they produce honey, pollen, propolis and wax, their most important activity is pollination, and all their other valuable products and additional benefits, such as the fact that I can spend hours admiring them and constantly learning from them and about them, come second. The current methods of agriculture pollute the environment, destroy forests and bee pastures by using chemicals and insecticides. It is life-threatening to bees, and therefore to us as well. If bees become extinct, there will be no one to pollinate plants, and since a significant portion of our food depends on pollination, humanity will quickly die out after the bees are gone. That is why it is extremely important that we value and protect bees and develop bee-friendly agriculture as soon as possible. Otherwise, there will be no bees left to tell anything to.

The Old Beekeeper's Smoker

I picked it up where he had dropped it,
Now almost rusted through
And blackened by many nights,
Since he had left it behind.

I sanded and polished it.
I repainted it silver.
I realized I had made a mistake
and sanded away my fresh flaw.

I began again, polished it again
three times in a row.
Finally, I oiled it carefully,
and when the snow came,

I lit it with a piece of fragrant bark.
He had told his children
How he had herded the bees
into their beautiful green hives.

I went out to them and said:
"Your father is dead, but don't cry.
I will be your father and continue
where he left off."

Diósd-Szentendre, February 4-9, 2021

Aristotle and the Observation of Bees

1

In the fall of 2022, I had the honor of being invited to a poetry festival in Patras, Greece's third largest city, and then to Athens. I was invited to both places because I had edited and written the foreword to the anthology *Young Hungarian Poets*, which was published in Greek and Hungarian. This week-long trip was a great opportunity not only to get to know contemporary Greek poets and poetry, but also to learn a little about the world-famous Greek beekeeping industry, whose roots, like everything else in Hellas, date back to ancient times.

Aristotle and animals in an 18th-century drawing.

I always get up at the break of dawn every morning, and I did the same when I was in Athens, so I arrived at the Acropolis just as it opened. It was raining, and there was almost no one there. I bought a combined ticket, allowing me to visit all the monuments in the area, then I climbed the slippery marble steps. I had the opportunity to contemplate, take Polaroid pictures, and write poetry, since I was almost alone, but as soon as the rain stopped, huge throngs of tourists rushed to the monument, and I thought it best to make a run for it. After the Theatre of Dionysus and the Roman Forum, I finally reached the ancient agora, where Socrates once made his contemporaries uncomfortable with his strange questions. The legendary Socrates, who believed that philosophy was a way of life and who wrote nothing but only talked and asked questions. His disciple was Plato, who founded his own academy and who already thought of philosophy somewhat as a business and a means of gaining influence; the sons of wealthy Athenian citizens came to study with him. Aristotle became Plato's student and an indispensable and influential figure in apiculture, as we shall see, with all his truths and errors. He founded an academy even larger than that of his master, and there is no doubt that, through his student Alexander the Great, he brought the influence of philosophy to its fullest expression in the Western world and history.

But the three ancient Greek philosophers mentioned above are also important from the point of view of European scholarship because they laid the foundations for the criteria of scientific observation and thinking. This is especially true of Aristotle, who lived on the island of Lesbos between 344 and 342 BC. There he wrote *Historia Animalium* which laid the foundations of zoology. For us, the fifth and ninth books are particularly noteworthy because they deal mainly with bees and beekeeping. With all their virtues and errors, they are the first surviving recorded scientific observations of bees in human history, and because they were written by Aristotle himself, they had an enormous impact. I should note that the tenth book, especially in beekeeping literature, should be considered "pseudo-Aristotle," meaning that it may have been written by one of his students, not by him.

The bee goddess from Rhodes.

It is no coincidence that in ancient Greece, even before Aristotle, bees must have been held in high esteem, and that depictions of a bee goddess from Rhodes date back two and a half thousand years. In one of the seven wonders of the ancient world, the temple of Artemis in Ephesus, the goddess Artemis was depicted with the eggs of wild bees, and she herself, the "mother earth", was described as half woman, half bee.

Aristotle's practical observations are, however, significant, interesting and suggestive of a surprisingly conscious and vibrant beekeeping tradition. For example, he describes how domestic bees are loyal to certain flowers, which allows beekeepers to produce honey of a particular variety. This floral fidelity means that the bee very rarely forages for a type of flower different from the one it started with, and so is site-faithful, meaning that as long as it finds flowers, pollen and nectar, it will return to the same area and to the same type of plant. Aristotle's observation was later studied and described by Charles Darwin. Aristotle observed the honey harvest, when beekeepers take the honeycomb, and then described how, in the autumn, beekeepers feed the bees sweet foods, figs and the like, to prepare for winter, Unfortunately, it is not specified exactly how this was done, but it is likely that a syrup was made from high-sugar fruits such as figs.

The Ephesian depiction of the goddess Artemis from a Roman copy.

One of Aristotle's most interesting observations was that bees "dance", what we now call the *waggle dance*, and it was only in 1788 that a German priest observed this more closely when he made a glass-walled hive. But the real significance of the bee dance was revealed by the Austrian ethologist Karl von Frisch, who spent decades researching and conducting crucial experiments, for which he was awarded the Nobel Prize in Medicine and Physiology which he shared with Konrad Lorenz. He described how the bee dance is truly a communicative act by which bees let their mates know what and where they can find outside the hive, and in this sense, a very specific and prominent form of social behavior.

In his *The Life of Animals*, al-Damiri Kamal al-Din asserted that Aristotle must have had a glass-walled observation hive, but al-Damiri died in 1405, long after Aristotle, who had never mentioned such a thing. The Roman, Pliny the Elder, was greatly influenced by Aristotle, and in his *Naturalis historia*, made two references to hives with walls "covered with transparent stone" through which the work of bees could be seen. There are several arguments to suggest that Aristotle himself may have built such a hive through which he could observe the life of bees. It's most likely, however, that he worked with beekeepers during his time in Lesbos, and they probably used horizontal, cylindrical hives rather than hives as we know them today. This has been confirmed by archaeological excavations. It is also possible that bees were kept in hive stones in the Mediterranean area at this time, as is also suggested by many traditions.

Aristotle was the first to observe and write that strong smells affect bees, that they attack people who are sweaty or highly perfumed, and that when a bee stings someone, it loses its stinger and dies soon afterwards. He described the brood, the swarming, the hexagonal shape of the cells and the use of smoke, its role, effect and importance in beekeeping. He even explained how bees can sense and predict the approach of storms and rain. As a practicing beekeeper myself, I have been confronted with the fact that this is why beekeeping just before a storm is not a good idea.

The Ephesian hive from antiquity

2

Aristotle described it, and I don't even rule out the possibility that he himself named *propolis*. What does propolis mean? The Greek is pro (π*ρο*) "in order, before" and polis (πό*λις*) "city". Before the city, or for the sake of the city, i.e., something before the *polis*, something for the *polis*. It is sometimes called bee glue by beekeepers, but most people know it as propolis. The main reason is that bees very often use this special and, to many, still mysterious resinous substance not only to cover the gaps in the hive but also around the entrance. But the term "polis" for Aristotle did not simply mean "city", but the highest and best opportunity for human society, existence, and even the fulfilment of human character. It is therefore significant that this term appears here and is still used in apiculture, and indeed in all languages, to mean the same thing.

Beekeepers mostly collect propolis in autumn, scraping it from various places in the hive or using propolis collection screens, and then extracting what is really valuable from the dry, hard and sticky material with an alcoholic solution. Propolis is a wonderful and extraordinary medicine, with antibacterial and antiviral properties, composed of at least 200 different substances. As the temperature inside the hive is constantly around 35°C, the humidity is high and there are large numbers of insects, the conditions would be ideal for a variety of pathogens, bacteria, viruses and fungi. The bees protect themselves against these with propolis. Propolis is used by bees as a building material, insulation, protection and adhesive. It is used to fill gaps and cracks to keep the hive hermetically sealed. The medicinal properties of propolis have been known to mankind for thousands of years and it has been used both internally and externally for its wound-healing, anti-inflammatory and antibiotic ingredients.

Aristotle called propolis “the tears of trees” because bees collect it from the buds, young branches and leaf stalks of trees when it is not nectar-gathering time. Its color is mostly brown, but it can also be yellow, green, red, black or white, depending on the plant from which it is collected. He had already highlighted its medicinal value, but for him the most interesting thing about bees was what was behind the propolis, the polis itself, the fact that bees are “state-building animals”.

Aristotle’s theoretical observations weren’t so much directed to bees as to what distinguishes humans from animals. What makes an animal an animal and a human a human? Which animals are similar to humans? Which animals possess distinct human characteristics, such as speech, organization, state formation? By the latter he meant, in particular, the *polis*, which exists “by nature (φύσει)”. In his study, Attila Simon discusses at length the importance of language, communication, community and politics in Aristotle, concluding that Aristotle considers only wasps, ants, hornets and, most significantly, bees to be similar to man in terms of their social organization. However, Aristotle is unwavering in asserting that all animals, including bees, are inferior to man

"...precisely because of the lower degree of openness of their communication, and indirectly because in animals collective action follows an invariant genetically coded pattern. The structures of biological activities and communicative capacities are naturally linked."
(Attila Simon, Language, communication, community and politics in Aristotle. In Other Minds. L'Harmattan, 2015.)

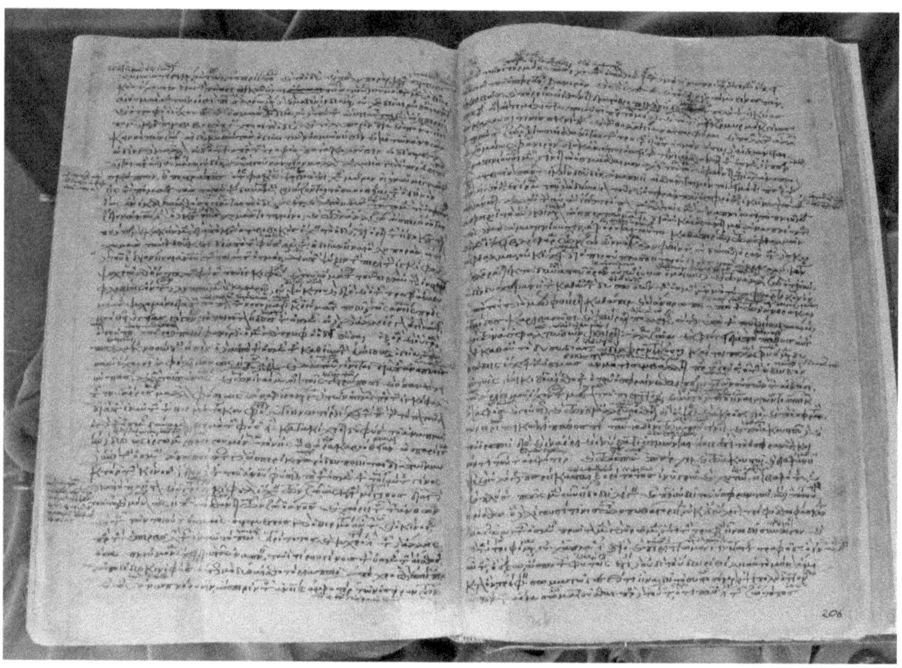

Copy of Aristotle's Historia animalium from Constantinople, 12th century (Biblioteca Medicea Laurenziana).

Aristotle himself writes: " Now, that it is self-evident that man is more of a political animal than bees or any other gregarious animals. Nature, as we often say, makes nothing in vain, and man is the only animal whom she has endowed with the gift of speech. It is true that sound is the sign of pain and pleasure, and therefore the other living creatures have it (for their nature has reached the point of having a sense by which they feel pain and pleasure, and they signify these to each other. Speech (λόγος), on the other hand, is used to make clear what is useful and what is harmful, and thus what is just and what is unjust."

But Aristotle was not at all at ease with the state-making nature of bees, and eventually came to the conclusion, especially because he could not satisfactorily decipher their sexuality, or more precisely their method of reproduction, that bees are "sacred beings". Daryn Lehoux writes about this at length, suffice it to say for now:

> "Indeed, Aristotle makes a comparable claim to divinity about no other animal except humans anywhere in his corpus."
> *(Daryn Lehoux, Why Does Aristotle Think Bees Are Divine? Proportion, Triplicity, and Order in the Natural World. Published online by Cambridge University Press, 2019.)*

3

And here we come to the errors of Aristotle. For example, as nice as it sounds to say

> "Honey falls from the air, principally at the rising of the stars, and when the rainbow rests upon the earth... "
> *(Historia animalium V.22.554a),*

we all know that it is actually the nectar of the flowers that the bees collect in their honey stomachs, and then take it to the hive, where, through a long process of digestion together, the nectar is ripened into honey, the water is evaporated and finally covered or sealed with beeswax.

A late edition of Mendez's book.

But the biggest mistakes by this first major observer of bees concerned his failure to understand how bees reproduce, and the way he distinguished four types of hive bees: workers, drones, little black thieves, and finally he called the mother (or queen) a king, mistaking her for a male. These errors had a huge impact on the history of European culture, since, as I mentioned, Aristotle described bees as "ideal", "polis", or state-creating and "divine" creatures, and these images have forever accompanied the history of humanity's relationship with bees. His teachings on bees were taken up by Pliny the Elder (23-79 AD) and Avicenna (Ibn Sina, 980-937), who all repeated and spread both his great insights and his serious errors. They and their successors compared ideal kingdoms to a beehive ruled by the king, unaware that the king is in fact a female bee. Indeed, even the Catholic Church was compared to a large beehive, in which the Pope himself was the king bee, and the worker bees were the priests, and what is more, they even made a beehive-like papal hat, which was in use for quite a long time.

It was Luis Mendez de Torres, almost two thousand years after Aristotle, who, in 1586, first tried to put an end to this misconception when he wrote in his book, with considerable humor

"I have seen with my own eyes that the king is laying eggs", in other words, that the king was in fact a female. Unfortunately, Mendez's humor was not much appreciated in the Kingdom of Spain, just as in the Kingdom of France most such attempts at bee science ended in contempt, or worse, execution, as bees were a model for a perfect society. And what if there is not a king at the center?

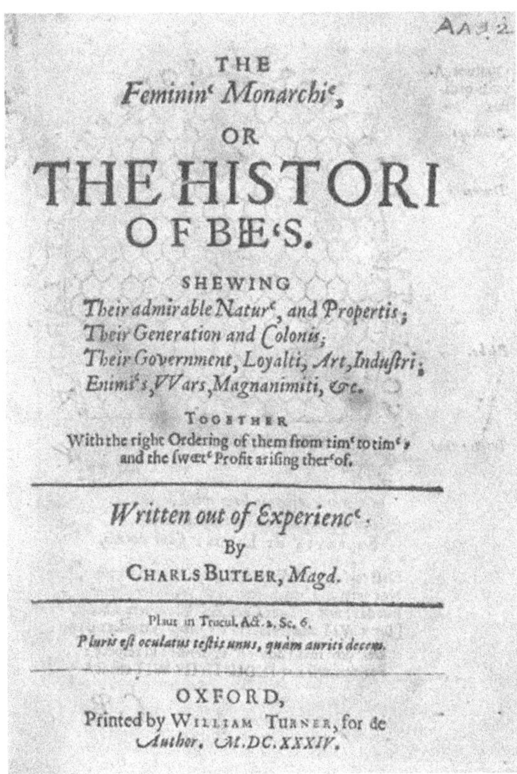
THE
Feminin' Monarchi',
OR
THE HISTORI
OF BEE'S.
SHEWING
Their admirable Natur', and Propertis;
Their Generation and Colonis;
Their Government, Loyalti, Art, Industri;
Enimi's, VVars, Magnanimiti, &c.
TOGETHER
With the right Ordering of them from tim' to tim':
and the sweet' Profit arising ther'of.

Written out of Experienc'.
By
CHARLS BUTLER, *Magd.*

Plaut. in Trucul. Act. 2. Sc. 6.
Pluris est oculatus testis unus, quam auriti decem.

OXFORD,
Printed by WILLIAM TURNER, for de
Author. M.DC.XXXIV.

Cover of Charles Butler's book.

Finally, in typical fashion, during the reign of Queen Elizabeth I in England, Charles Butler described in his 1609 book *The Feminine Monarchie* that the king is in fact: the queen. Incidentally, Butler continued to rely on Aristotle's authority by deliberately, if erroneously, translating king as queen, thus repairing his eminent predecessor's greatest error by creating another. Of course, the Queen of England was not angry that that there was a queen at the center of what she thought was a perfect society of bees, and Charles Butler's book was a great success. Butler also projected his own social classes onto the bee society, assuming princesses and princes, a working class and a military. Butler even composed a madrigal from the quacking of the queen bees, but we will discuss this later in the discussion of the relationship between bees and music.

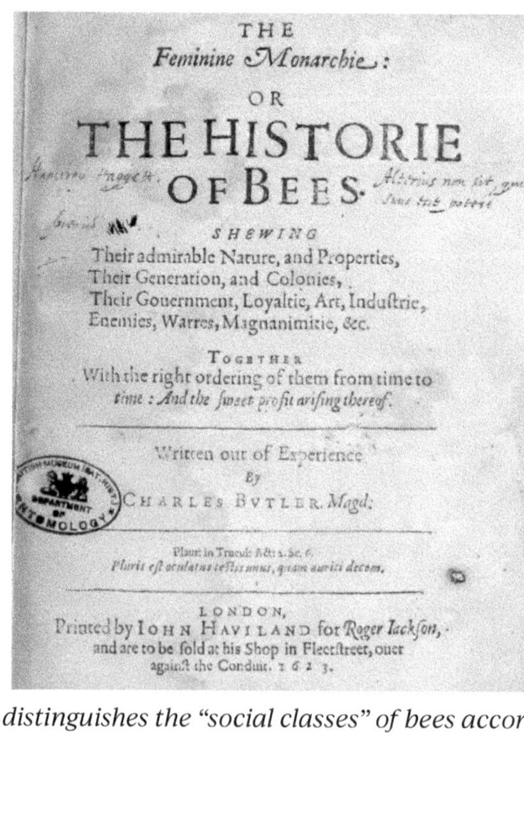

THE
Feminine Monarchie:
OR
THE HISTORIE
OF BEES.
SHEWING
Their admirable Nature, and Properties,
Their Generation, and Colonies,
Their Gouernment, Loyaltie, Art, Induſtrie,
Enemies, Warres, Magnanimitie, &c.

TOGETHER
With the right ordering of them from time to
time: And the ſweet profit ariſing thereof.

Written out of Experience
By
CHARLES BVTLER. *Magd:*

Plaut. in Trucul. Act. 2. Sc. 6.
Pluris eſt oculatus teſtis unus, quam auriti decem.

LONDON,
Printed by IOHN HAVILAND for *Roger Iackſon,*
and are to be ſold at his Shop in Fleetſtreet, ouer
againſt the Conduit. 1623.

Butler's book distinguishes the "social classes" of bees according to their age.

After Charles Butler, the father of English beekeeping, we should also note the name of Richard Renman, who in 1637 discovered that worker bees are also females. Traditionally, in Hungarian beekeeping, we call the only mature, mated and ovipositing female, the center of the colony, not the queen, but the mother. One might probably draw some far-reaching philosophical conclusions from this about Hungarian beekeeping particularly since Hungarian apiaries have traditionally been family businesses for a very long time.

Aristotle's observations on bees reflect his own philosophy and thus himself, his virtues and his faults. This is another fascinating example of the interaction between observer and object of observation. Perhaps this is why I hear reports from my beekeeper friends about why beekeeping goes badly if you are in a bad mood. Or, conversely, how much the bees helped them when they were sad? I'm not sure about this, but the fact is that the scientific observation and description of bees began with Aristotle. Since then, observers have gone beyond his observations on countless points, including, for example, understanding the learning and communicative abilities of bees and discovering that our little friends are capable of much more than we previously thought.

These were the things I pondered as I left the ruins of Aristotle's academy behind me, and then visited a small beekeeper's shop, where I had a long chat with a very nice Greek lady beekeeper. I found out that they mainly uses tackable beehives and she found my horizontal beehives very strange and unique when I showed her pictures of them. What was interesting to me was that on one of the Greek islands, Leros, beekeeping is used not only for honey production and propolis extraction, but also for psychotherapy. They use beekeeping to help people with psychological problems, with mental illnesses, for whom working with bees can be healing. This was particularly fascinating because the famous American poet Sylvia Plath also suffered from severe depression and, in the latter part of her life, tried to cure herself by beekeeping. (I will write about that in my next essay.) As we were saying goodbye, the Greek beekeeper divulged the recipe for one of the oldest and most delicious Greek sweets, the sesame honey bar, which was mentioned by Herodotus and is said to be 6,000 years old. It is also called *pasteli*, although it is probably an Italian name, but in Crete, for example, it is called *koptoplakous* or *gastris*. Similar sweets can be found all over the Middle East, even in India (rewri) and Vietnam (Kẹo Gương), but if it is so ancient, it has had time to spread, just like Aristotle's ideas.

Finally, I managed to get four types of typical Greek honey. On my return home from my adventures, my wife and I enjoyed tasting them and comparing them with our own honeys. The Greeks are convinced that they have the finest honey in the world, and when I tasted them there, I had the strange feeling I had to agree with them; but tasting them on my return home, I thought our chestnut linden honey was better than any of them. I've come to strongly suspect that everything is best and tastiest where it's produced, and so the best honey is that which comes from the local producer - in Greece from Greek beekeepers, in Hungary from Hungarians, and here at home, undoubtedly our own.

Bees and Souls or from Sylvia Plath's Bees to Beekeeping that Heals the World

1

Winter is a special time in the life of bees and beekeepers. I am often asked, "Is this the time when bees sleep?" - and people are always surprised when I tell them what happens to bees in winter. Bees settle in for the winter only after the tidying and thinning of their nests, light wrapping or insulating of the hives, mite monitoring and control, and repairing the lattices made to keep mice out in the autumn. In the cold, they form a cluster in the center of the hive's combs, then they slowly move from front to back on the combs, consuming the overwintering food in the honeycombs. They are in a cluster with the central figure of the hive, the queen bee, in the middle, I tell those asking, and with their tiny shivering movements, kinetic energy, keep the family warm. If the queen is sick or injured, if the bees' fat bodies are not properly developed, i.e., if we don't have many healthy, fat, furry bees, if the overwintering food is insufficient or not well stocked, if pests - mites, mice, birds - bother the bees, if the bees' environment is noisy, the bees have a harder time overwintering, with greater losses in the spring. But if all is well, from the autumn equinox onwards, the bees start to look inwards, entering a strange state, which reaches its low point at the winter solstice, when there is complete peace and quiet around the hive. Then the queen starts to brood again, the colony slowly starts to move, and by the spring equinox, we're able to witness a wonderful outburst of activity. Of course, in many cases the weather and its changes may have interfered, and what was once unthinkable in our part of the world is now more common, namely that bees do not have this resting phase, but continue to swarm throughout the winter. On several occasions I have witnessed how, as the weather cools down around Christmas, the bees' clusters break up and fly out, and when the weather suddenly cools down, they have to work very hard to grow clusters again. But the natural rhythm of our bees is that they move in during the winter, swarm out in the spring, develop, reproduce, gather in the summer, settle in the autumn and start all over again, according to the solstices, equinoxes and weather. This is the cycle to which good beekeepers must adapt; they must listen to and live with their bees, while at the same time seeing the cycle of nature at both micro and macro levels. Such a thorough, detailed, and in-depth activity not only reinforces external observations, but also directs attention inwards. A good beekeeper not only works outside the beehive, not only inside the workshop, but also inside himself. Beekeeping is, in a good way, not only external work, not only animal husbandry, not only honey production, but also a spiritual and mental process. Good beekeepers are concerned about their hives, but they know that the laws of nature are strict and that they can, if necessary, break up one hive to save three others.

The author's apiary in winter.

Good beekeepers will aim to keep their colonies strong and healthy, taking only the surplus honey. Good beekeepers know that if their bees are doing well, they themselves will also do well. A good beekeeper comes into direct contact with nature's most remarkable creature: bees. This intimate relationship between human and insect has inspired thousands and thousands of myths and stories, which I have tried, and continue to try, to explore and reveal in *The Art of Beekeeping*. Now I have come to one of the most extraordinary and most difficult to relate: the relationship between bees and the human soul. Through two stories, I want to show you exactly what it's all about, hoping that my readers, beekeepers or not, will be intrigued to delve deeper, and be inspired by what they find.

It's a well-known fact that honey and propolis have a beneficial effect on human health, but wax is also used in beauty care. The air emitted by bees, bee venom and even the buzzing or vibration of the bees themselves have also been used in apitherapeutic houses, since they have been found to have, in many cases, a very powerful healing effect. The use of bees and bee products for healing is known collectively as apitherapy. The medicinal properties of honey are well known and have been used by mankind since ancient times, just as the antibacterial and antiviral properties of propolis are now familiar to all. Less is known about bee venom therapy, although it has been used to great effect in patients with rheumatism and sclerosis. The filtered air released by bees has been used effectively in treating those suffering from asthma. There are few things more soothing than relaxing in a bee house, or as it is more fashionably called today, an "apitherapy house", with bees buzzing around.

What we beekeepers take for granted and don't even notice is a miracle to most people. But let's just think about how, during the season after having worked the whole day with the bees and we lay our heads down in the evening, we can still "hear" the buzzing, and our lungs are filled with the sweet smell as we fall soundly and peacefully asleep. It's no coincidence that bees are considered idyllic beings, even though we know that the world of bees is rigorous and logical. I have seen more than one excellent beekeeper, even after many years, marvel at the way a hive lives, organizes itself, pulsates. Sometimes when I'm tired, I sit on a folding chair and watch the bees dart in and out of the hives in a golden stream. Strangely enough, I'm always instantly relaxed and at ease, and somehow have a feeling of "everything is fine". I've often meditated among my bees, during the workday, or on the occasional morning, and as my mind became more relaxed, the bees showed themselves more. They would land on my hand, or I would suddenly notice how the bees would surround and trap a hornet which we had fought together over the summer. Nature really shows itself when you don't just watch it, but really observe it, when you pick up its rhythm. Picking up the rhythm of our bees means that in winter we ourselves are also drawing in, taking stock and preparing for the coming year. The great masters of Hungarian beekeeping who were here before us - Ambrózy, Örösi and Zsidei - considered winter to be a very important time when beekeepers can study, read, rest and prepare. So beekeeping also has a certain spiritual effect, which has been observed and studied by people much cleverer and more knowledgeable than I.

2

Sylvia Plath.

I have already referred to Sylvia Plath, an outstanding 20th-century poet. Plath was born in the United States, lost her father in early childhood, and as a brilliantly talented poet, was awarded a scholarship to England, where she met Ted Hughes, an exceptional English poet. They married and had two children together. In the meantime, they moved to the English countryside, and in a particularly poignant radio interview they talk at length about it. And, of course, they both wrote beautiful and moving poems. However, Sylvia Plath suffered from a very serious mental illness, from depression, and this is reflected not only in her poetry but also in her most famous work, *The Glass Bell Jar*. A mother of two, she was unable to cope with her role as a housewife and mother and suffered increasingly serious episodes. Her self-analysis, typical of poets, soon revealed that she had been "not okay" since the untimely loss of her father, and several of her poems are about "the father" and her ambivalent relationship with him.

> "Sylvia Plath was eight years old when her father died, so she doesn't have many memories of him. Yet, as time went on, he endowed her with more and more mystical qualities. In the end, she saw him, or his death, as the cause of all her failures. His father's absence was present throughout his life."
> *(Kinga Fabó, Schizophrenia and language. A personality sketch of Sylvia Plath. 77)*

In the aforementioned interview, the poetess said, "I think I was happy up to the age of about nine - very carefree - and I believed in magic..." Sylvia Plath's father was Otto Plath, himself a beekeeper, who wrote a very valuable book on bumblebees. Bumblebees are so interesting and important that I would like to write about them in a separate book. Coming back to Plath, György Gömöri, a literary scholar and poet living in England, writes:

The cover of Sylvia's father's book about bumblebees by Otto Plath

"The role of bees in both Plath's and Hughes' poetry deserves a special analysis. (...) Hughes unpacks the (beekeeping) episode from Sylvia's stylization and shows how beekeeping is related to the Father, who emerges as a ghost from the well of the unconscious. Of course, Hughes could have guessed this earlier, as he obviously knew that Otto Plath was an entomologist, had written a book about bees, and Sylvia

> describes herself as THE BEEKEEPER'S DAUGHTER in her very first volume."
> *(György Gömöri, I remember your fingers. On the poems of Ted Hughes. Holmi, 1998/6, 853)*

On June 8-9, 1962, Sylvia Plath and Ted Hughes attended a beekeepers' meeting, which Sylvia wrote about at length in her diary, noting that she was their leader:

> "He indicated a pile of hives, like white and green blocks of wood with little gables & said we could have one, if we would like to fix it up."
> *(Sylvia Plath, Diaries. 1950-1962, Anchor Books, 660)*

The next day they took the hive home, with the bees inside, and this started a very important, instinctive process in Plath's poetry: she began to write "beekeeper poems". Plath's "beekeeper poems" were more about her inner, spiritual struggles than about bees, as we read in the notes to the Hungarian edition:

> "Plath holds the beekeeper poems in high esteem, planning to place them at the end of the Ariel volume. The subject matter is rich: his father published a scientific treatise on bumblebees, and honey may be a reference to his severe diabetes. She herself, as a formidable but mortally wounded queen bee, is a third giant to the two ghostly figures, the father and the 'other'."
> *(Poems by Sylvia Plath. Európa. Budapest: 2002, 225)*

If you want to read these poems in Hungarian, you have to get both published Plath volumes, and they don't have all of her poems either. Of Plath's beekeeping poems, one of my favorites is *The Arrival of the Bee Box*, in which the poet writes:

I wonder if they would forget me
If I just undid the locks and stood back and turned into a tree.
There is the laburnum, its blond colonnades,
And the petticoats of the cherry.

They might ignore me immediately
In my moon suit and funeral veil.
I am no source of honey
So why should they turn on me?

Tomorrow I will be sweet God,
I will set them free.
The box is only temporary.
(The Arrival of the Bee Box. Sylvia Plath: Ariel. Selected poems)

In this beautiful poem, and in the few lines quoted, it is clear that for the poetess, working with bees was a multifaceted activity. It was not simply a hobby. Plath's husband Hughes, also a poet, wrote in another poem

"When you wanted bees I never dreamed

It meant your Daddy had come up out of the well."

- and it was probably precisely her relationship with her father, or lack of it, that Plath was reliving in her relationship with bees, since her father was also a beekeeper. Plath was therefore probably the first to use beekeeping as therapy, but this therapy was started too late and self-therapy has its limitations. Among Plath's last poems are the bee poems, because although she had fed the bees, she herself could no longer overcome her depression and on February 11, 1963, at the age of thirty, a mother of two, she locked herself in the kitchen and turned on the gas. She committed suicide.

Every human being is a deep, deep well.

The cover of Sylvia Plath's poetry collection Ariel, which was published after her suicide and brought her world fame (A bee in the left corner)

3

Since then, there have been many literary studies on the relationship between Sylvia Plath and bees, but I want to draw attention to the fact that beekeeping can also be a kind of self-knowledge, which the sensitive poetess instinctively realized, but which she could not unravel because she was overcome by her mental illness.

Much later, however, in modern Greece, in Leros, beekeeping was used in psychotherapy. In my previous essay, I mentioned that the name Leros had come up during my travels in Greece, as a place where beekeeping has been used to help those suffering from depression and other psychological and mental disorders for many years. The history of the Leros asylum in the 20th century is shockingly tragic, since it was first as an Italian military fortress, then a re-education camp for Communist youth, then a prison camp, and later a very notorious asylum. I asked my friend and correspondent Marianne Lipcsei, who lives in Rhodes, to dig a little deeper into the subject, and I quote from her letter:

> “The disclosure of the inhumane treatment of inmates in the asylum founded here in 1957 has gone down in history as a great scandal in 1989. That year, the world became aware of what was happening in Leros thanks to the journalist John Merrit, who reported on his visit there on the front page of *The Observer*. As he put it, the gateway to the underworld in ancient Greece was in a cave in the Peloponnese, but in modern Greece the gateway to hell is on Leros. Under pressure from the ensuing international outcry, the Greek government finally took measures to end the situation in Leros and introduce a nationwide psychiatric reform. In Leros, the aim was to eradicate the inhumane treatment of patients and to rehabilitate them socially and occupationally.

This was given a legal framework in 2002 with the registration of the Social Cooperation of the Dodecanese Department of Mental Health, which provides employment in its Caserma Botanical Farm for the limited number of patients at the psychiatric institute still operating in Leros, helping them to reintegrate into society. And if I'm not mistaken, they started beekeeping in 2015 for that purpose. The history of the asylum is in itself an endlessly shocking phenomenon. Two documentaries have been made about it and are available online. Unfortunately, I have watched them... It is hard to believe that the peaceful, beekeeping patients are now on the same Leros where these films were shot a few decades ago. What sounds really fantastic is that they are employing these patients full-time (i.e., paying them for therapeutic beekeeping), and some of them are now living in independent housing rather than in psychiatric wards."

It seems that beekeeping in Leros and the Caserma Botanical Farm have been a real innovation.

Leros therapy has become very famous, resulting in several reports and many articles on the subject. It is worth emphasizing that the patients work together there with expert psychologists, psychiatrists, therapists, so they are not trying to heal themselves alone, and thus beekeeping is just a "pretext" in a good sense, through which they can reconnect with nature, themselves, others and possibly with normality.

Psychiatric patients are beekeeping on the island of Leros

Closely related to this, a major study was done in England in recent years, which is probably one of the most remarkable papers on the subject. In it, Lindsay Cordery-Bruce has written in detail about her many years of experience in a small apiary, working with people suffering from various addictions, mainly alcohol and drugs. She reports the results and - more importantly - how they were achieved. She describes how she talked to patients about the activities and work in the apiary throughout the year, and how the life and hard work with the bees helped them to recognize and move on from their addiction. This study is very well documented and very important if bees are to be used to heal people's souls in the *future* (see *Lindsay Cordery-Bruce: Cross-pollinating the evidence: A feasibility study into the therapeutic potential of beekeeping for adults with substance misuse issues. Cardiff Metropolitan University. 2021*).

Bee therapy for mental illness is nothing new in modern society; as early as in the period after World War I, it was observed that soldiers with post-traumatic stress disorder who returned to America after the war were able to settle back into their lives more quickly and easily if they worked with bees. Hence, beekeeping was used for soldiers with similar problems. But in an important and extensively researched publication on this topic, entitled "Can beekeeping improve mental health in times of crisis?", the focus is on more recent times, in particular the Covid epidemic with its wide psychological impact. The authors found that those who lived in the countryside and had the opportunity to spend more time in nature during the quarantines were better able to cope with the quite severe psychological and mental effects of the pandemic and the resulting isolation. In their study they describe how

> "it is not just individuals who are appreciating how beekeeping can help with anxiety and other mental health issues. Beekeeping as a therapy has been used for a number of years, for instance as an intervention to dissuade young offenders from reoffending (Tierney, 2012) and in cases such as Post Traumatic Stress Disorder (PTSD) in military veterans (Learning English, 2019)."
> *(Karin Alton and Francis Ratnieks, Can Beekeeping Improve Mental Wellbeing during Times of Crisis? Bee World. 2021. 99:2. 43.)*

They add that whether you're walking in the woods, spending time in your garden or watching bees, research suggests that just two hours a week of these activities, regularly, of course, can help most people to combat anxiety and maintain mental health. Of course, in more severe cases, more serious and supervised therapy led by an expert therapist may be needed. So, beekeeping, gardening, walking in the woods will not solve all the problems of modern mankind, but it can help us to maintain mental health, to regulate our relationship with nature. Let us not forget that, however much we may think that this is not the case, we are part of nature.

The art of beekeeping, or apiculture, thus leads to ecotherapy, or healing with nature, which is a direct path to biophilia, or love of nature. This triangle does not solve the increasingly catastrophic global crisis unfolding around us, but it can go a long way to helping us find the right answers to the questions that plague us and to remain normal people in an abnormal world. Bees can teach us about introspection, solidarity, moderation, self-sacrifice, adaptation, survival, as well as about outflow, diligence, love of nature, true abundance and real happiness. The greatest can always learn the most from the least.

Bees and Outer Space, or Will There be "Space Bees"?

When I asked a friend, "Do you think bees have ever been to space?" he laughed at first, then when he saw me smiling conspicuously, he became suspicious and asked back, "Tibor, are you saying that there are 'space bees'?"

No, there are no "space bees" – at least not as far as we know at the moment – but the truth is that there have been more bees in space than humans. Roughly 3,420 bees, which is a surprising number compared to the nearly 600 humans who will have left our planet by the end of 2022 and have ventured in space. Anyone dismissing this question at first glance lacks the curiosity, imagination or receptiveness to serious questioning or scientific observation, without which not only life but also conscious beekeeping would be very difficult to develop. The study of bees in outer space and the results obtained bring with them some very thought-provoking conclusions for our work and cooperation with bees on earth.

When I tell lay people about the world of insects, or show them the life of bees up close, or even take a frame out of a hive with bees on it, and explain the biology of bees, I have heard many times from many people that "they are aliens". But that's nonsense; bees are very much terrestrial, in fact, according to the current state of science, they are more ancient than we humans, having evolved many millions of years before homo sapiens. The bees, like humans, are a species that has significantly altered the surface of the planet. By favoring pollination by insects rather than self-fertilization or accidental wind pollination, bees encouraged plants to produce more flowers. As this pollination proved to be more successful, more flowering plants were added to the planet's surface, and species that did not favor pollinating insects were pushed to the background. A regular competition for bees among plants to develop more colorful, attractive flowers rich in pollen and nectar began. The result of this interaction, which has lasted for millions of years, are colorful spring meadows, fragrant linden groves, dangling acacia blossoms, nectar and ultimately the honey on our table. So, bees are intimately linked to the planet, just as the whole flora is linked to the fauna and vice versa by a million times a million strands. That's because the nature that surrounds us – of which we ourselves are part, even if we tend to forget it – is an interwoven system. The plant and animal worlds are inextricably linked, and when we subtract an element or add in an inappropriate one, the whole system collapses like dominos. Insects, bees and other pollinators are particularly important and sensitive elements in this system, and their disappearance or endangerment has fatal consequences for the whole ecosystem, including humans. I cannot fail to recommend the books on this subject by our contemporary David Goulson, in particular *Silent Earth - Averting the insect apocalypse (London 2021).* But if you want to read about the current state of insects in English, I recommend free downloadable Insect Atlas, published in 2022. *(Related link: https://mtvsz.hu/rovaratlasz2022).*

It is estimated that humanity would survive the extinction of pollinating insects for only four years; the famines we would face immediately would only be the first of many apocalyptic consequences. Do we really want to see this?

What will happen to bees in the future? Is the solution, as billionaire entrepreneurs and phantasmagorists claim, that bees will leave the dying planet with us? Or something else entirely? What does science say and what does literature say? Perhaps it is precisely because of these questions that it is no wonder that pollinators, including our honeybees, have been a very prominent feature of human history from the very beginning, as I have shown in previous writings on the art of beekeeping. For this reason, bees could not have been left out of the conquest of space, that is, space travel.

Space bee in the author's hand, Jánossomorja, 2023.

The first bees went into space on board the space shuttle Columbia between 22 and 30 March 1982, as part of a NASA experiment devised by American high school students. The bees were provided by the Coplin Bee Farms in Alvin, Texas, and the results of the study were published by Todd Nelson and James Peterson in a paper entitled "Insect Flight Observation at Zero Gravity". To be precise, they took on board 50 different insects - bees, butterflies, flies, several of which were in the pupae stage. Among the insects travelling into space were 14 honeybees. The experiment proved disastrous – at least for the bees – but very instructive for us. The flies that were "born" in space adapted relatively quickly to the microgravity conditions, while the adult moths and adult worker bees had serious navigational problems. The butterflies flapped their wings, while the honeybees simply did not fly, but rather clung to the side of the observation box, and if they lost their grip, they "just floated hopelessly and desperately" in the air. The biggest problem, however, was that even though a feeder was placed in the observation box, the bees, separated from their hive and the first to go into space, refused to take any food. So, one by one they had starved to death by the end of the journey. It is reasonable to ask whether the simulated environment, the zero-gravity situation or the separation from the hive caused the first "space bees" to die of starvation.

In 1984, a second, much more serious experiment took place on board the spacecraft Discovery, when a whole colony of bees was finally taken on board, including the queen bee, a total of 3,400 bees, in a special observation hive with frames, a small space for flying and a feeder. The aim of the study was to observe honeybee behavior, survival and wax formation under microgravity conditions. The crew took notes throughout the eight-day mission, which once again provide some remarkable lessons to be learned.

On April 6, 1984, nine hours after the lift-off, the bees had apparently survived the launch. On April 9, video footage was taken of the bees' behavior, their short flights and their clinging together on the wall of the test crate. On April 11, they were again filmed and on April 13, after a thorough inspection, the astronauts found that the bees had – surprisingly and unlike on the first space flight – fully adapted to the gravity-free space and had learned to fly from one point to another. They fed and even reinforced their hive frames with wax, albeit in a slightly different way than they do on Earth, since without gravity it is difficult to tell up from down. In all, 120 bees died, which seems perfectly normal for such a short period of eight days. Furthermore, the queen laid a total of 35 eggs. This is what strikes me about this experiment, that the queen laid so few eggs relative to the time available, as under ideal circumstances she would be able to lay many, many more, even thousands of eggs a day. It is likely that the total of eight days of the experiment was not enough time for the bees to adapt in all respects, but the low mortality and flight adaptation were still remarkable compared to the first trip, I repeat: in eight days. Just think of the long and exhaustive training astronauts receive to do their job in space, where everything is completely different from here on Earth. The bees had to adapt from their natural position, without any preparation, and on the second attempt they seem to have succeeded, which is no mean feat, but the drastic drop in brooding may well be a sad conclusion. After all, a worker bee lives for about 15-38 days in season – and 150-200 days during overwintering – but even with our current technology it would take years to travel to the nearest planet, Mars. So, a non-reproducing colony of bees would very likely be depleted and die during the first phase of the journey. And it is very difficult to create a self-sustaining space colony that can survive for many years without the fruits and vegetables that grow there being contaminated. But even if we were to find some way to get the bees to the space colony, for example by hibernating them, how would they behave once they woke up?

Special beehive on the 1984 space mission. Source NASA.

The answer to this question was sought at the experimental laboratory, LunAres, set up in a nuclear bunker in Pila, Poland, where a lunar and a Martian space base were simulated and then colonized by 90,000 “bee space cadets”. The aim was to see if the bees could join a possible mission to the Moon or Mars. To study the effect of the confined space on honeybee colonies, the researchers ran two two-week experiments, *Habeetat-1* and *Habeetat-2*.

Logo for Habeetat-2. Source LunAres.

The hive was a tent sealed from the inside, where scientists continuously monitored the temperature and humidity of the hives, the variation in the weight of the bee colony, among many other things. Lavender, pollen, water and one or two artificial light sources were placed next to the hives inside the tent. The study was published at the International Astronautical Congress. The bees were studied in a microgravity environment, as I wrote above, but they did not look at how a whole colony of bees behaves in isolated conditions, said Aleksander Waśniowski, one of the leaders of the experiment. He went on to say

"Around 1,000 to 1,200 bees died every four days. It was almost a little scary because you don't see this in nature, but in isolation the dead bees covered the floor. In a normal, garden-variety hive, it's not unusual for hundreds of bees to die per day. A honeybee queen can produce well over a thousand offspring per day, more than enough to replenish the dead and even grow the colony."

A bee colony in isolation during the Habeetat-2 experiment. Source LunAres

This is something we beekeepers can observe up close in the spring, but Waśniowski and his colleagues suddenly noticed that the queen of the isolated hive had stopped laying eggs, the temperature of the hive had dropped, leading to the death of more bees. A few bees did fly out for water and a small cluster formed around the artificial lights, but they completely ignored the lavender they had been given, as well as the pollen. Until the surviving and remaining members of the hive settled into a "defensive posture", or what we would call an overwintering cluster, exactly as they would when they set up to survive the cold months. So, the conclusion is that in isolated conditions, bees first suffer severe losses and then become full and passive. It will therefore be very difficult to get them to do any pollination on an isolated lunar or Martian base. But let us ask ourselves, are humans ready for years of space travel and then living on a relatively confined space base? Not just physically, but even psychologically? If the series of shutdowns around the pandemic has taught many people anything, it is that, like our bees, we are social creatures. So I think it unlikely that, with our current technical capabilities, bees could leave the Earth for any length of time. We may not be able to do so either. Just think: even ignoring the spiritual aspects, how much are human muscles affected by the absence of gravity for even a very short period of time. All this points to the need to save what we can here on Earth.

Special beehive on the 1984 space mission. Source NASA.

With all this in mind, it is worth drawing my readers' attention to a related work of fiction, namely Maja Lunde's *The History of Bees*, which is really more of a young adult novel, half-science fiction. The story, which takes place in three timelines, is about the extinction of bees. Although it is perhaps not the best novel I have ever read, I can recommend it, because it is so thought-provoking. It describes a not-too-distant future – but what is in some parts of the world, unfortunately, a terrifying present – when people will have to pollinate flowers themselves in order to have a crop to eat. In this book we read a phrase that no beekeeper should ever forget, namely:

"Bees cannot be tamed. They can only be bred and cared for."

Preferably here on Earth, because it seems that it won't work elsewhere, because there really are no such things as "space bees", I hasten to add.

Music of the Bees

1

During the winter, a good beekeeper's only tasks are preparing for the coming spring and regularly going out to the hive to tune in and listen. A low, steady buzzing sound means that the hive is wintering well. Loud buzzing indicates a serious problem, such as mice or shrews disturbing the overwintering bees, or the death of the queen bee. The latter is what beekeepers call crying bees. If there's a resounding silence in the hive, that's the worst. It means the bees are dead.

I used to attach a vacuum cleaner hose to the exits and listen for bees; then I borrowed my doctor neighbor's phonendoscope, but not long ago we developed a small electric device with a sensitive microphone that makes it easy to listen through the thin board walls of hives. I can even connect a small tape recorder to it so I can pick up the buzz. As I step up to the hives on a foggy winter morning and put my ear to each hive, listening for a long time to the deep and monotonous buzzing, I find myself thinking about the music of the bees. I wonder, as I listen to the hum of the overwintering cluster, how these wonderful creatures live in such a sensual presence. Bees have their own unmistakable music, from which we can learn a lot. I remember the first time I dared to touch a bee without gloves, bare-handed, I wrote the following poem:

The Beekeeper's First Bee

Today I caught a bee:
I think it might have been my first.
The first bee I held in my bare palm,
the first from the hive, the first
bee of my life. A tiny, vibrating being,
messenger of the Sun, friend of the flower,
a wonderful vegan! Creating life from all its life,
and today I took this fragment
of a swarm-soul in my palm,
and it began to dance in circles!
It danced away the sunlight,
the golden autumn fields, my garden,
and in my garden, the hive and beekeeper,
the two of us!
The inside of the hive, the combs
filled for winter, the pollen, the brood, and
what about the mother?!
This little bee sang with dance and danced with song,
and when I carefully put it back
at the entrance to the hive, it turned around,
flew around my head, and
and buzzed to me: „Oh, beekeeper,
it is not I who sing, it is not I who dance!
Oh, beekeeper, it is the cosmos that
sings while dancing and dances while singing;
oh, beekeeper, never forget."

Szentendre, October 8, 2020

Then, in the spring of 2022, I visited 22 kindergartens in Hungary as Mr. Beekeeper in an educational puppet show called *Bees are the Essence*, written by Csilla Gévai. I noticed then how much the children loved to imitate the buzzing of the bees. This is the first thing that people remember about bees, besides the more standard yellow and black portrayal of their appearance.

Our author visited 22 kindergartens as a beekeeper in the spring of 2022, performing in Bees are the Essence.

This is the buzz.

The music of the bees.

Bees were once thought to be deaf; there is now, however, serious literature on the "music" of bees, which do not simply buzz, but vary their buzzing, making different sounds in different situations (or rather, vibrating at different frequencies). As I wrote above, they make one sound when they are hibernating peacefully and another when they are in trouble. But they make a different sound yet again when they are sick or when they are about to swarm. Queen bees also have their own sound, different from the others. In Hungarian, this is called quacking, a word that imitates the sound. I have already written in connection with ancient Egyptian beekeeping *(see the essay "Daughters of the Sun"*) that, by imitating the quacking of the queen bees, some sources claim that Egyptian priest-beekeepers were able to "sing" the mothers out of the hive to facilitating the swarming.

Paying attention to the sound of bees, drawing conclusions from it, even imitating it, have probably all been an important part of studying bees from the very beginning. More modern electronic monitoring devices, which are inserted into the hive, track not only temperature and humidity but also the bees' sounds, and alert the beekeeper to what is happening or what is being prepared in the hive, based on the different vibrations. With the help of my cousin Pityu, who is an electrical engineer, I developed three similar devices, which we called BeePityu, based on his expertise and my ideas, and we then aggregated the resulting data on a computer.

The spring placement of our self-developed BeePityu into the hive.

In addition to temperature and humidity, we also collected data on the bees' buzzing and observed its changes.

This method still needs a lot of refinement, but as we become more adept at processing data, and as modern technology becomes less expensive, I think there is a chance that soon you will be able to simply hold your phone up to the hive entrance and the program will tell you what's going on inside, just by the buzzing of the bees - or more precisely, the frequencies of the buzzing. This way, it might not even be necessary to put separate devices in each hive, but rather just one sensitive microphone and a program with big and reliable data that can independently identify the sounds. In Slovenia I have seen software and hardware similar to ours used to monitor bees; we are still in the tinkering stage, but if I had the capital and a year to spare, I would most certainly develop such a program and device. Of course, all this is only a dream at the moment. Why would such a program and device for monitoring bee music be a good thing? Because in beekeeping, the point is to touch our bees as infrequently as possible and to avoid constantly opening the hive. In fact, to do otherwise, we often risk harming and disturbing the bees. With the least interference possible, we can realize our goal of maintaining the healthiest colony producing, for itself and us, the most excellent yields possible.

Observing the music of bees is essential for a good beekeeper. When I go out to work with the bees, I first hear their buzzing, and the next thing that strikes me is the smell of the hive. It's that subtle, sweet, unique cloud of scent that is unmistakable, even as it mixes with the pleasant smoke coming from my smoker.

I go to the banks of the Danube, looking for fallen, old, rotten, powdery willows, which provide the best smoker material. I stuff it into my smoker; it burns slowly, giving a pleasant smoke which is not intrusive and soothes the bees. This smoke does not overpower the smell of bees and comb. One could discuss the smell of the bees separately, the fact that the queen bee keeps the colony together with her own pheromone, and that each bee may not only have an individual sense of vision but probably also a specific sense of smell, which allows them to sense when their own queen or another queen is nearby. While their own queen will be fed for many years, if you put a foreign queen among them without any warning, she could easily be killed. True, I have seen the opposite happen before my eyes, when I simply allowed a new queen onto the comb of a queenless family. There are two possible scenarios in such cases; one is that the new queen has an attractive smell (i.e., pheromone) to the queenless hive and/or, the second, as a good friend of mine put it, "the bees have not read the literature which says this is impossible". The dance of the bees is now a well-known behavior. We know that it is their way of informing each other of nectar and pollen collection sites, but it is also their way of showing where it is worthwhile to swarm, and at the same time the bees make different sounds, buzzing, as they say in common parlance.

This buzz is unmistakable, unique, repetitive, and often changing. It is interesting that when, for example, the sound of a colony of bees preparing to swarm was recorded and played back to another colony, it did not swarm. Hence, it is more likely that bees are processing complex information, i.e., I think they are communicating with smells, vibrations (buzzing) and dance (movement) at the same time. I'm almost certain that these are inseparable, so you can fool the bees separately, for example, you can use scent to make them swarm where you put the scent – and there are tools for that – but I think a simpler way, prevention based on observation, is always more effective. A good beekeeper knows exactly when his colonies are swarming and what is causing it: they may be short of space, they may have a lot of honey, and so on. But swarming is bees' natural breeding process that may possibly be built into future beekeeping, even if most beekeepers would like to avoid it for now, since the swarming family takes the bulk of the honey, and not many beekeepers like to chase the swarms up trees.

However, if we pay attention to the music of bees, we can learn from it, and if we have the right tools for data collection and data processing, we will sooner or later know in advance, without opening the hives, which family is swarming. In other words, we will understand the language of bees.

Will we ourselves be able to send them a message, if we understand which vibrations and scents mean what and can reproduce them, or is this just a poetic fantasy? The human imagination is limitless, and behind every great discovery is a dream. What was thought unimaginable yesterday is now part of everyday practice. Art and science have always inspired each other, and this is particularly true in the 21st century, where environmental crisis and technological revolution are both on the rise. Can our digital technological achievements help solve the ecological crisis? For example, an artificial intelligence program that can detect not only from the buzzing of a bee colony whether it is about to swarm or is queenless, but also whether it is sick or poisoned, offer a wealth of opportunities for prevention and thus for protecting bee colonies.

2

But *The Art of Beekeeping* is about much more than just practical observations. After all, theory and practice, art and science are closely linked, and we can hardly progress in life if we do not seek to see the whole picture by looking at the details.

When we look at a single bee, we must see the whole colony behind it.

Bees are social creatures, living in a complex system, and they have fascinated humanity from the beginning, and for good reason. In understanding insects as superorganisms, man has always wanted to understand a little of himself, his own place in creation; this is true for why people observe any part of nature with so much sensitivity and intelligence.

When the kindergartners dressed up in bee costumes and came in as a group in preparation for our performance, they started buzzing spontaneously. People love to imitate, and imitation is particularly key to children's learning process. As I listened to the buzzing children, I wondered how the buzzing of bees in could be echoed in our music.

I have already mentioned in previous writings the seminal book by the influential English beekeeper Charles Butler (1560-1647), *The Feminine Monarchie, or The Historie of Bees*. Butler was not only a beekeeper, but also a musician, and composed a special madrigal from the bees' song. *Melissomelos*, or *The Bee's Madrigal*, is a choral work that does not use the bees' irregular buzzing but rather imitates the queen bee's quacking – known in English as piping – as the hive prepares to swarm. In his book, Butler included sheet music for the madrigal, so it is still performed today. It is truly wonderful music, and I have had the pleasure of hearing it.

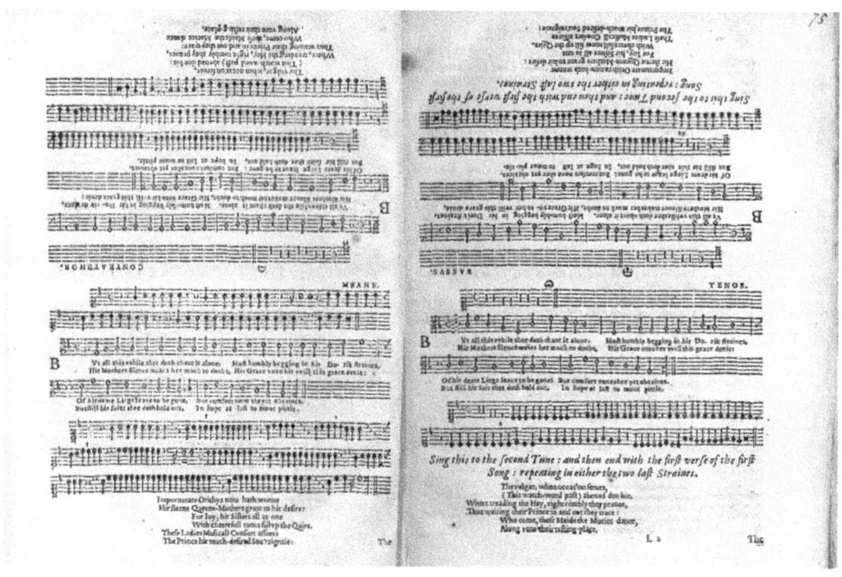

Charles Butler's Melissomelos, or The Bees Madrigal from his book The Feminine Monarchie, or The Historie of Bees (London, 1609).

Both classical and contemporary composers have been interested in the buzzing of bees. Perhaps the best known is the German composer François Schubert – not to be confused with Franz Schubert, who died at a tragically young age! – who wrote the bagatelle for violin and piano, *L'abeille*, or *The Bee*.

I think that if they were to listen, many would also recognize Chopin's piano etude *The Bees (Etude No. 14, F Minor, Op. 25, No. 2)*. If one would like to refresh one's musical memory, I would personally recommend Artur Rubinstein's performance of this piece.

But what everyone knows – if from nothing else, then from films – is Nikolai Andreyevich Rimsky-Korsakov's *Flight of the Bumble Bee*, which I recommend listening to performed by Yehudi Menuhin and Gerald Moore. Few people know that this work is part of a larger opera, *The Tale of Tsar Saltan*, written for the centenary of Pushkin's birth and first performed in 1900. In it, Rimsky-Korsakov, in homage to Pushkin, adapted one of his tales into an opera, so we know for a fact that *The Flight of the Bumblebee* was inspired by a passage from a Pushkin poem in which the prince is transformed into a bumblebee and can fly anywhere. This is how the poem is translated in English,:

> In a trice, he shrank, and lo!
> Ere or he could even gasp,
> He had turned into a wasp.
> Then he buzzed, and rapidly
> Overtook the ship at sea;
> Gently settled aft, and stole
> Out of sight into a hole.

(Translated by Louis Zelikoff)

In the music of the bees, buzzing, poetry and music come together in many ways. To give you more than just classical examples, there is Gloria Gaynor's hit *Honeybee*. But there is also the theme song from the film *The Secret Life of Bees*, screened at the Bee Film Festival. The song is sung when a lady beekeeper is buried after her tragic suicide:

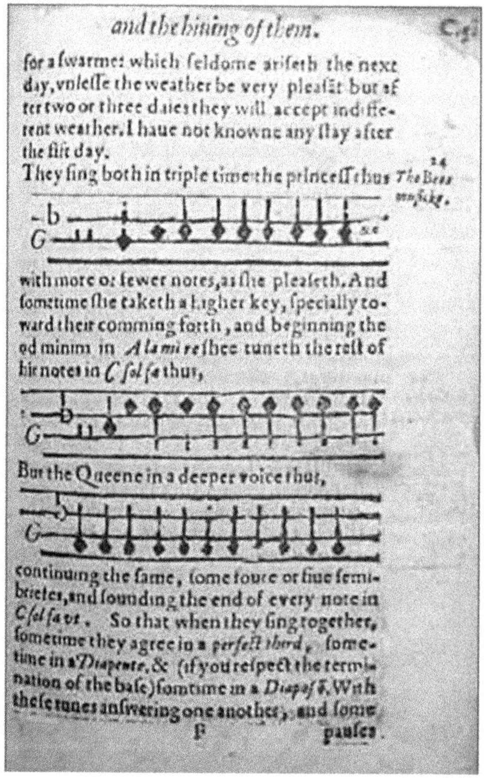

and the hiuing of them. C.3

for a swarme: which seldome ariseth the next day, vnlesse the weather be very pleasãt but af ter two or three daies they will accept indiffe- rent weather. I haue not knowne any stay after the fift day.

They sing both in triple time: the princesse thus 24 *The Bees musicke.*

with more or fewer notes, as she pleaseth. And sometime she taketh a higher key, specially to- ward their comming forth, and beginning the od minim in *A la mi re* shee tuneth the rest of hir notes in *C sol fa* thus,

But the Queene in a deeper voice thus,

continuing the same, some foure or fiue semi- brieses, and sounding the end of euery note in *C sol fa vt*. So that when they sing together, sometime they agree in a *perfect third*, some- time in a *Diapente*, & (if you respect the termi- nation of the base) somtime in a *Diapas* &. With these tunes answering one another, and some

F pauses.

A transcription of the "quacking" of the queen bee from the book by Charles Butler (London, 1609).

"Place a beehive on my grave
And let the honey soak through,
When I'm dead and gone
That's what I want from you.
The streets of heaven are gold and sunny,
But I'll stick with my plot and a pot of honey.
Place a beehive on my grave
And let the honey soak through."

This leads to the topic of my next essay, but we cannot ignore the song written for the film score by the contemporary American composer Eliot Goldenthal. In 2010, Goldenthal wrote a great score for the film of William Shakespeare's play The Tempest - which I think is better than the film itself - and whose song *Where the Bee Sucks* is originally a short excerpt from the play (V. 1), sung by Ariel to Prospero, which sounds like this in English:

Where the bee sucks, there suck I,
In a cowslip's bell I lie;
There I couch when owls do cry.
On the bat's back I do fly
After summer merrily.
Merrily, merrily, shall I live now,
Under the blossom that hangs on the bough.

It is particularly interesting how well Goldenthal contrasts text and music to create the atmosphere of Shakespeare's drama, at once sad, mystical and beautiful. We have seen how the music of the bees is not only interesting from a practical point of view, but also how it has inspired and enriched human music with valuable musical creations throughout history.

Shakespeare, moreover, referred to bees several times in his plays and drew very serious and far-reaching conclusions for human society. Some have suggested that he was a beekeeper himself, as well as a dramatist, but this is perhaps less important than what it meant to project the life of bees onto human society through the ages, or more precisely to anthropomorphize bees, to endow bees with a human character. But I will write more about this in my next essay on the relationship and interaction between bees and humans.

Bees and Man

1

I usually look forward to the first spring inspection. But before, I keep wondering: did my bees overwinter well? Did I do the right things last the autumn? Did I prepare the bees well enough for the cold temperatures of the winter dormant period?

The beekeeper's first and biggest test occurs when the temperature rises to 15 degrees C (59 degrees F) and he can finally check his hives. Though this used to happen in early March, thanks to milder winters now, it can take place as early as February. When I open a hive for the first time, I feel joy and excitement, just like the very first time I ever opened a hive. From my first mentor, my Uncle Misi, I learned to make decisive, quick and purposeful movements in order to get a clear picture of the family as quickly and with as little disturbance as possible.

The strange and reciprocal relationship between the observer and the observed is something I remember from the first time I did this. Since poetry also requires precise observation, and poetry is very often an attempt to capture the interplay between the microcosm and the macrocosm, and since I have practiced Buddhist vipassana meditation a lot, in my more sensitive moments I pay attention not only to what I observe, but also to how and in what way the experience of observation is recorded in me. The intense practice of beekeeping has led me to notice how my observation affects the observed, i.e., my bees. The stronger and more concentrated my presence, the faster I can get the picture, the less I disturb the bees, the more effective I am. I'm almost certain that it is possible to reach a point in working with bees where I become so at one with them that I can even work without any protective equipment. I admit I have not yet reached that level, but I have seen a beekeeper who has, who lit the smoker just for the atmosphere and left it at the end of the hive, forgetting to lower the veil of the beekeeper's hat, just using it as a visor. He made no unnecessary movements as he listened to the assemblage of bees, to their sound, and as he sensed their smell. When he took out a frame, he did not even accidentally crush a single bee, and it was as if the bees showed him what he needed to see. He immediately saw if there was pollen, honey, brooding, and immediately spotted which frame the queen was on. Such a beekeeper is a role model for me, although I myself can only occasionally work with my bees in what modern psychology calls being "in the flow". I use the smoker, and after a few good stings in the face, I don't really feel like taking the hat off anymore.

With our author's mentor, "Uncle Misi" - Mihály Hudek - in 2021 in the Szentendre apiary.

Our author after a few stings in the face. :)

On the other hand, I'm used to beekeeping without gloves most of the time. With my hands bare, I have a better feel for where I am and get instant feedback if I make a mistake. This is sometimes a painful experience and resulted in 148 stings last season. These are 148 good reasons to pay even more attention to my work, to bother my bees even less, and when I do check on them, to do so with the utmost thoughtfulness: quickly, efficiently, accurately. Being stung is a part of beekeeping, an inconvenience that can be tolerated - if you're not allergic - but it has been particularly motivating for me. I took it as a message from my bees so that we were able to work together in a great way and have stronger, healthier families because of it. It was a tragedy when my hives died at the end of one wintering, six in a row, but I didn't cry, I cleaned up the dead, burned them so they wouldn't infect other families, and the bad experience spurred me on to learn more, to work better.

I was reminded of this when I travelled to Java for the third time in 2013 and visited the rice and tea fields on the huge, 3,000-meter-high volcano Gunung Lawu, the "Mount Lava". At dawn, I noticed how a tiny old man drove two giant buffalos to pull his plow through the rice fields. He had been working all day, but he noticed I was watching him all the time, and after work he came to see me, and we sat down for tea. The old man was almost a hundred years old, spoke a special Javanese dialect, so we needed an interpreter. I asked him at length how he could be in such good physical condition at such an age? He answered all my questions, including this one, with a serene smile, saying that he was not in good shape, that he was half blind, that his old bones ached, that he slept a lot, but that he was proud that he and his second wife, thirty years younger than him, still got on well. We laughed a lot, and then I said, "But I've just watched you plowing the rice field from early morning until late afternoon with two huge buffalos, which takes some serious stamina." The old man kept smiling and said that the world was made up of five elements: water, earth, air, fire and soul. We can always experience these elements, even in ourselves. The right thing to do is to strive for balance. For example, the element of fire often manifests in humans as anger. When he looked out over his rice fields at dawn that day, he became extremely angry, because of his four children, twelve grandchildren, and who knows how many great-grandchildren, none of them were there to keep up the rice fields. He felt great anger welling up inside him, but he thought how much he loved his children, grandchildren, and great-grandchildren and could not be angry with them; they are simply other people with different lives. They have to realize for themselves how important farming is. So, he took that anger and turned it into a force within himself that helped him through the hard physical work of the day. In this way, he worked easily, hurt no one, and restored the balance of the five elements within himself.

Perhaps I am not exaggerating when I say that my conversation with that hundred-year-old Javanese man on Mount Lava changed my life. I began to observe the five elements in myself and in the world. The excesses and deficiencies of the elements, their effects and consequences. It didn't matter whether the teaching of the five elements was true; what was more important was that it worked. Striving for harmony is a fruitful process in itself, and observing the elements encourages me to turn the excesses, evils, and deficiencies in the world and in myself into good. If nothing else, it is enough to observe and realize again and again that evil also has a place in the world.

At the same time, I was never – nor am I now – a passive observer at all, but I am very much engaged in observation, and I am aware that what is outside also appears inside, and what happens inside also has visible signs outside.

I cannot always influence events, and in fact, the more I practice this path, the more I realize that the influence of an individual is quite small, but at the same time, this smallness can change the world. That is why I have imagined two piles, one containing the phenomena and events that I cannot change, and the other containing those that are the way they are because of me. For example, the first pile includes terrible and depressing things such as epidemics and war, while the second includes my own work, my writings, my relationships with my family and friends, my behavior, my actions, and my thoughts. Although the latter may also have aspects that I have no influence over, I can sometimes influence things that I thought were impossible to change, so there is an intersection between the two sets. This intersection is life and death itself. There is no doubt that we will all die, myself included, but until then we must live. It is up to us how we spend our lives: do we treat others with love or hatred? Do we learn, experience, or do nothing?

I would like to note that in a world overly focused on production and revenue, it is difficult to explain how important it is sometimes in life to do nothing, or more precisely, not to interfere, and how effective this can be in some cases. For example, novice beekeepers typically open their hives every day, unaware of how much damage they are causing. Sometimes it is better not to interfere, but just to observe. At the same time, bees are semi-domesticated animals, which means that without human intervention, not only would they not produce honey, but they would probably die out, except for a few more resistant colonies, just as most of our pets would if we did not take care of them.

The agricultural revolution is estimated to have taken place twelve thousand years ago, so it might be more accurate to adjust our current world's calendar to that, although there is no doubt that the teachings of Jesus, which appeared two thousand years ago, are at least as important from a spiritual and human perspective. Ultimately, it doesn't matter what year we are in; it may be much more important where we are in our own lives.

So, following the teachings of the Javanese man's five elements, I can easily recognize when one of my bee colonies is sick or weak, or when there is trouble in the apiary and I feel anger rising within me.

Why is this so? What did I do wrong? What could be the problem?

I take this anger and transform it into a willingness to act. I examine the bee colonies and try to do what I do better and more skillfully each time. And when a bee stings me because I accidentally strike it while removing a frame, I immediately transform the pain and anger into even greater attention so that I make fewer mistakes. This attitude has a long tradition in human society, history, and mythology, and Tomáš Sedláček gives countless examples of this in *The Economics of Good and Evil*. In this wonderful book, he draws attention – following Mandeville's *The Fable of the Bees* – to the fact that evil also has a place in the human world and is even necessary in shaping it. This brings us slowly to one of our most important topics, the relationship between bees and humans, bee society and human society, their parallels and, I dare say, their interaction.

2

We have already mentioned in connection with the music of bees (see *The Music of Bees*) that there are numerous references in Shakespeare's plays to suggest that the great poet must have had some relationship to beekeeping. In his interesting essay, Richard Grinnell went so far as to suggest that, in addition to writing plays, the poet himself kept bees on his country estate *(Richard Grinnell, Shakespeare's Keeping of Bees. ISLE: Interdisciplinary Studies in Literature and Environment. 23/4. 2016, 835-854).*

Who has ever heard of such a thing?

How could poets be involved in beekeeping? Yet Virgil wrote at length about bees in *Georgics*, and Sylvia Plath also dedicated several beautiful poems to them (see *Bees and Souls*), in which both drew on their own experiences. It is quite possible that Shakespeare also kept bees, although this is impossible to prove, as is the true identity of the great playwright, as Bill Bryson points out in his entertaining book on Shakespeare:

> "Although he left nearly a million words of text, we have just fourteen words in his own hand – his name signed six times and the words 'by me' on his will. Not a single note or letter or page of manuscript survives . . . We have no written

> description of him penned in his own lifetime. The first textual portrait – 'he was a handsome, well-shap't man: very good company, and of a very readie and pleasant smooth witt" – was written sixty-four years after his death by a man, John Aubrey, who was born ten years after that death."
> *Bill Bryson: Shakespeare: The World as Stage*

Bryson, an excellent essayist, like most Shakespeare scholars, goes on to write at length about how strange, mysterious, and inaccessible the world's most famous playwright is to posterity, and how, before he became a famous writer during his lifetime – especially between 1585 and 1592 – we simply do not know what he did or where he was. Some say he was a schoolmaster in the countryside (perhaps he was a beekeeper?), others saw him serving in the military in Italy or Flanders, while still others believed that he took to the sea and sailed on the Golden Hind alongside the legendary pirate Sir Francis Drake. All these legends are based on the well-known uncertainty surrounding his whereabouts and the sources of his outstanding knowledge and expertise.

> "It is often noted, for instance, that Shakespeare's plays are full of ocean metaphors ("take arms against a sea of troubles," "an ocean of salt tears," "wild sea of my conscience") and that every one of his plays has at least one reference to the sea in it somewhere. But the idea that this argues for a maritime spell in his life shrivels slightly when you realize that *sailor* appears just four times in his work and *seamen* only twice. (...) Shakespeare's marine allusions mostly depict the sea as a hostile and forbidding environment, a place of storms and shipwrecks and unsettling depths – precisely the perspective one would expect from someone who wasn't comfortably acquainted with it."
> *(Bryson, p. 57)*

With this in mind, we read the great writer's beekeeping metaphors somewhat differently and are less inclined to assume that he himself was a beekeeper. The truth is, however, that this is less important than what Shakespeare puts into the mouth of the Archbishop of Canterbury in his play Henry V.

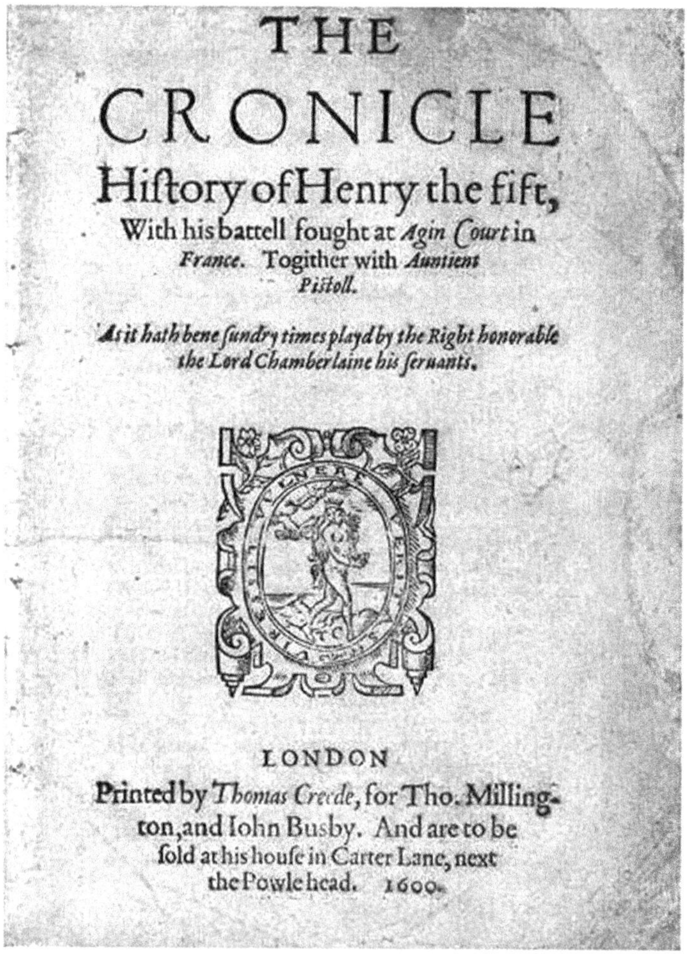

THE

CRONICLE

Hiſtory of Henry the fift,

With his battell fought at *Agin Court* in *France*. Togither with *Auntient Piſtoll*.

As it hath bene ſundry times playd by the Right honorable the Lord Chamberlaine his ſeruants.

LONDON

Printed by *Thomas Creede*, for Tho. Millington, and Iohn Busby. And are to be ſold at his houſe in Carter Lane, next the Powle head. 1600.

Title page of Henry V.

Let us recall the relevant part of the drama, when the archbishop convinces King Henry to go to war against France with the following monologue:

BISHOP OF CANTERBURY

Therefore doth heaven divide
The state of man in divers functions,
Setting endeavor in continual motion,
To which is fixèd as an aim or butt
Obedience; for so work the honeybees,
Creatures that by a rule in nature teach
The act of order to a peopled kingdom.
They have a king and officers of sorts,
Where some like magistrates correct at home,
Others like merchants venture trade abroad,
Others like soldiers armèd in their stings
Make boot upon the summer's velvet buds,
Which pillage they with merry march bring home
To the tent royal of their emperor,
Who, busied in his majesty, surveys
The singing masons building roofs of gold,
The civil citizens kneading up the honey,
The poor mechanic porters crowding in
Their heavy burdens at his narrow gate,
The sad-eyed justice with his surly hum
Delivering o'er to executors pale
The lazy yawning drone. I this infer:
That many things, having full reference
To one consent, may work contrariously,
As many arrows loosèd several ways
Come to one mark, as many ways meet in one town,
As many fresh streams meet in one salt sea,
As many lines close in the dial's center,
So may a thousand actions, once afoot,
End in one purpose and be all well borne
Without defeat. Therefore to France, my liege!

Divide your happy England into four,
Whereof take you one quarter into France,
And you withal shall make all Gallia shake.
If we, with thrice such powers left at home,
Cannot defend our own doors from the dog,
Let us be worried, and our nation lose
The name of hardiness and policy.
KING HENRY
Call in the messengers sent from the Dauphin.
Attendants exit.
Now are we well resolved, and by God's help
And yours, the noble sinews of our power,
France being ours, we'll bend it to our awe
Or break it all to pieces. Or there we'll sit,
Ruling in large and ample empery . . ."

(William Shakespeare, Henry V. Act I, lines 191-234)

What can we see in this short excerpt from the play? A rhetorical turn of phrase supported by a magnificent poetic image, which according to our current knowledge – is biologically nonsensical. However, people of Shakespeare's era saw their own society reflected in the society of bees, as has been the case throughout the ages until today. The pharaohs compared Egypt to a beehive, and Aristotle considered the bee colony to be a polis, describing its central figure as a king.

Later, it was only when Luis Mendez de Torres, published his book published in 1586 (see: *Aristotle and the Observation of Bees)* that a bee colony was imagined as a kingdom, and it was only very slowly that the idea that the king lays eggs, i.e., is female, entered the public consciousness. It was only after Charles Butler's book, published in 1609, that the king was recognized as a queen, a term still used today in many Indo-European languages. Thus, we continue to anthropomorphize the bee colony, even attributing to the central figure a role more characteristic of human kingdoms than of the world of bees. This is why I emphasized that Hungarian beekeepers call the queen "mother," which is much more accurate and apt in terms of her actual role. But the misunderstanding of bees as an ideal society is born of ignorance and, more often than not, intentionality. The attribution of human characteristics to the bees, is not a simple mistake, but is much more than that: it is an inspired misunderstanding that has opened up new possibilities.

When people have observed bees, they have often believed they were discovering themselves and their own society, or even an idealized society. They observed and watched this superorganism, interacting with it, just as the beekeeper and his bees interact. Continuing with this line of thought, it is well worth taking a closer look at a few examples of how the society of bees, their observation, description, misunderstanding, and understanding have influenced us, humans, our human society, and our history.

3

Bees have been considered sacred since ancient times, and their importance was probably recognized very early on. Honey, given that it is a complete food and does not spoil, it quickly became popular as a natural sweetener and was even deemed one of the foods of the gods. A reed and a bee appear in hieroglyphics ancient Egypt (see: *Daughters of the Sun*), and even Aristotle wrote with admiration about the polis of bees (see: *Aristotle and the Observation of Bees*), but, as mentioned, since he could not confidently determine the name of the queen bee, he called her king, which, with his "tiny" mistake, set off a whole avalanche in European cultural history.

Medieval European kingdoms. with their strict caste systems in which everyone had predetermined roles, were compared to the idealized society of bees, although, in fact, a bee participates in every phase of work during its lifetime. In his seminal book of 1609 *The Feminine Monarchie*, Charles Butler describes bee society in much the same way as Shakespeare describes contemporary England in *Henry V*, with its castes of nobles, soldiers, merchants, and workers; in Butler's version however, the queen bee has at last been recognized as a queen.

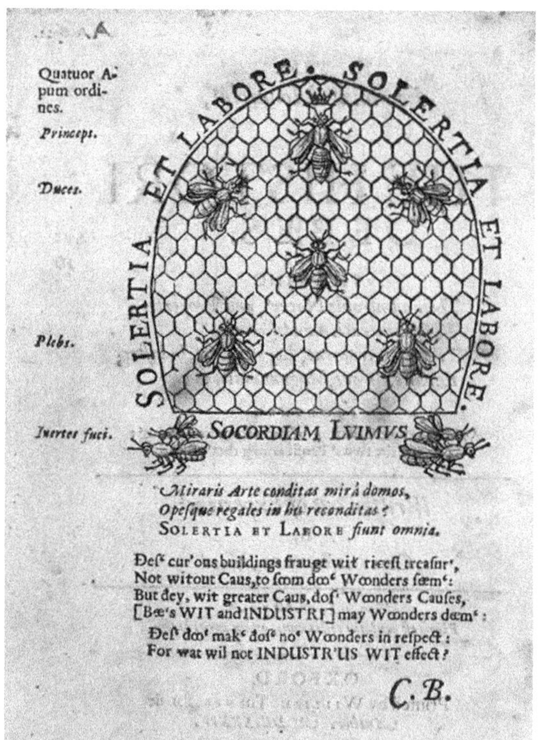

Charles Butler: From his book The Feminine Monarchie, a drawing dividing bee society into castes, as in Shakespeare's play Henry V.

Louis XII, King of France, described himself as Rex *non utitur aculeo*, meaning "a king without a stinger." The "perfect" society of bees, the pharaoh's empire in Egypt, the polis in Aristotle's time, and the absolute monarchy in the Middle Ages were seen as places where the "king" ruled, and his subjects worshipped only him. Countless emblems from the 16th, 17th, and even 18th centuries, on which we see beehives or bees, serve as political and moral support for the *ancien régime*; that is, they sought to idealize the period of absolute rule that preceded the French Revolution. If we think that this disappeared with the French Revolution, we are greatly mistaken! The French Revolution used bee symbols in the same way: the hexagon, the sign of a honeycomb cell, became the symbol of France, and in many places the Republic also adopted the symbol of the beehive. After the fall of the Revolution, Napoleon used the bee as his own imperial symbol, his cloak was embroidered with golden bees at his coronation as emperor, in order, apparently to unite the imperial hierarchy with republican ideals, at least on a symbolic level. After all this, it is not surprising that Freemasonry also adopted the beehive as its own symbol of great significance.

Humanity thus endowed the bee society, which they considered ideal from the beginning, with many more characteristics that were more applicable to themselves. This was particularly true of the ruling ideologies, as it proved very attractive, especially to the rulers of the time, that a society should exist as a superorganism for the sake of efficiency and that its members should carry out their assigned tasks without question. Hopefully, communism and fascism disillusioned humanity somewhat with this idealized image, this insect-like ideology. As Juan Antonio Ramírez writes in his fascinating book, which deals primarily with the relationship between bees and architecture from Gaudí to Le Corbusier in terms of bees and social metaphors:

> "To sum it up it can be said that the long period between the Napoleonic Wars and the Second World War perpetuated the metaphor of the beehive as the perfect united and industrious society. To be more precise, however, it was a

rather ambiguous political symbol, equally adaptable to the most peaceable ideologies and the most bellisoe and reactionary."
(Juan Antonio Ramírez: The Beehive Metaphor, From Gaudí to Le Corbusier, Reaktion Books, 1998, translated by Alexander R. Tulloch, p. 23.)

In his book *The Beehive Metaphor*, he adds a little later that:

"Such a vision is important because it can explain, partly, the proliferation of negative ideas that had been associated with the beehive since the Second World War. The idea of a people or nation that functions like a superorganism and before which personal individuality disappears was emphasized by the Nazis, but is not alien to anarchist and communist tradition (though for different reasons). The defeat of the fascist powers saw an attenuation of the 'positive' connotations associated with social insects, and very few people have dared to advocate the resurrection of the beehive as a symbol of political activity. In fact, I know of only one great politician of modern times who has done so, François Mitterand."
(Juan Antonio Ramírez: The Beehive Metaphor. From Gaudí to Le Corbusier. Reaktion Books. 1998, p. 24.)

Although it may seem to many that modern consumer society is capable of more than its predecessors, from Henry V to Hitler, there are those who believe that, with the support of mass media brainwashing, society will indeed be able to reduce skeptical, individualistic, thinking, people to units without skepticism or individual will that work until death like insects. The eminent Czech public intellectual Tomáš Sedláček had a different view, arguing that it was tempting for every tyranny to think of individuals as cogs in the social machine, to deprive them of their individual human nature, rights, and desires, to make them easily replaceable and thus dispensable, and to alienate them from their environment. However, seemingly inexplicable feelings such as friendship, compassion, and love have always resisted this transformation of people into insects, no matter how tempting this may be from the perspective of dehumanized power.

THE
FABLE
OF THE
BEES:
OR,
Private Vices,
Publick Benefits.

LONDON:
Printed for J. Roberts, near the *Oxford Arms* in *Warwick Lane*, 1714.

The cover of Bernard Mandeville's book The Fable of the Bees from 1714.

Even more interesting is how authors who endowed bees with human characteristics enriched our self-knowledge and social awareness with their unique observations. One of the most important authors in this regard is Bernard Mandeville, who reformed European public thinking by linking the concepts of private vices and public benefits in his book *The Fable of the Bees*, first published in 1714 and then in 1723. This book was at least as influential and dangerous as Machiavelli's, and anticipated Adam Smith's "invisible hand" theory of markets. Mandeville opens his book with a 433-line poem that guides us into a society of bees, the beehive, where all that is evil has come to prevail. The bees are selfish, egotistical, sinful, but suddenly the "bee god", hearing their outpouring of complaints, bestows infinite goodness upon them. The once prosperous bee family then becomes lazy and unmotivated, until they are defeated and destroyed, with only a small group remaining to swarm out and seek refuge in a nearby tree hollow.

Mandeville concludes his tale with the following words:
"Such were the blessings of that state;
Their crimes conspir'd to make them great

(...)
"Then leave complaints: fools only thrive
To make a great an honest hive.
T' enjoy the world's conveniences,
Be fam'd in war, yet live in ease,
Without great vices, is a vain
Eutopia seated in the brain.
Fraud, luxury, and pride mull live,
While we the benefits receive :
Hunger's a dreadful plague, no doubt
(...)
Bare virtue can't make nations live
In splendor ; they, that would revive
A golden age, must be as free,
For acorns as for honesty.
(Bernard Mandeville, The Fable of the Bees.)

Yes, Mandeville praises private vice as a necessary evil that constantly drives society forward. In Christian logic, from Jesus to

St. Thomas Aquinas, the existence of evil, wickedness, and the devil was explained as the driving force behind human creation, spurring it forward. As Tomáš Sedláček points out in his book, Mandeville honestly exposes evil and does not reject it but presents it as a socially useful phenomenon. I quote:

> "The thesis that partial evil contributes to the good of the whole and therefore is not advisable to remove is one we have repeatedly encountered in much older writings. (...) Jesus discourages his disciples from pulling out weeds, 'because while you are pulling the weeds, you may root up the wheat with them. Let both grow together until the harvest'. And Thomas Aquinas reminds us: 'Many goods are present in things which would not occur unless there were evils.' "It may have been unfortunate for Mandeville that he was not aware of these sources, because referring to them would have certainly saved him much of the controversy his fable provoked."
> *(Tomáš Sedláček, Economics of Good and Evil. 189)*

Mandeville's provocative social explanation, a distorted mirror in which he showed how hypocritical human society is, struck a chord with the sensibilities of the age – and perhaps still strikes a chord with us today – but was branded as corrupt, erroneous, and sinful. People saw him as the antichrist himself. David Hume and Jean-Jacques Rousseau were among his opponents, and he was universally reviled, from John Wesley to Adam Smith. It is true that the latter learned a great deal from him and subtly replaced "selfishness" with "self-interest" in his own book, which already sounds better. But it was Mandeville who compared bees to a far from perfect human society to illustrate that human society is fundamentally made up of sinful individuals whose private sins and vices actually function for the public good. Mandeville's book sometimes describes the beehive, as a way of looking critically at his own society, but elsewhere we see him describing an idealized bee society, in which bees are always endowed with human characteristics and humans are seen through the perceived or real perfection of bees.

In this context, we might not believe it, but now the most progressive and interesting contemporary research on bees concludes that bees' decisions and their society seem to be the most democratic.

4

The dance of bees, and dance in general as a means of communication in the world of humans and animals, deserves separate study here. For now, I will just touch on this topic to draw attention to the work of the legendary professor Karl von Frisch, who, while studying the dance of bees, found how bees communicate with each other. Karl von Frisch not only observed this but also proved in his book that after scout bees find a source of nectar, they return to the colony and dance to indicate the location of the source in relation to the hive and the sun. The other bees understand the dance, allowing all the foraging bees to find the same source without having to search for it.

The cover of Karl von Frisch's book from 1954 containing his Nobel Prize-winning discoveries and proofs.

Frisch's discovery, deciphering the dance of bees was truly revolutionary and conclusive, earning him the Nobel Prize in 1973. After World War II, Frisch reorganized the Zoological Institute at the University of Munich and continued his research. Martin Lindauer, who was primarily involved in behavioral science, joined him in this endeavor. Lindauer was undoubtedly one of Frisch's most important and talented colleagues, who, during their joint research, noticed some research bees that did not bring back any pollen or nectar and, unlike the others, were dirty, but danced as if they "wanted to say something to the others." As he made his observations, it became clear that the "filthy little bees" were actually looking for a new home for the swarm that was about to leave the hive, and since this was happening after the world war, they found suitable nests in the chimneys of numerous destroyed and abandoned houses in the area. Lindauer saw that the bees decided together on the time and place of swarming, but a truly accurate description of this was first presented in the book by contemporary researcher Thomas D. Seeley. Seeley wrote of Lindauer:

> "Lindauer"s pioneering investigation of the swarm bees' process of collective decision making provided just a preliminary understanding of the subject, and that it, like all good scientific studies, raised many more questions than it answered."

These studies asked such questions as how bees decided where to swarm when they left the hive, and "how can a small cluster of bees hanging from a tree branch make such a complex decision?" And finally, "how can a swirling group of ten thousand bees flying in the air steer themselves and stay together during their flight to their chosen home, which is typically a small hollow in an inconspicuous remote corner of a forest?" Lindauer's questions – and those of Seeley, who followed in his footsteps – are more than justified. Seeley attempted to explain these phenomena in his book, and sought answers to the following questions:

> "This book aims to present to biologists and general readers [...] about how the workers in a honeybee swarm conduct a democratic decision-making process to make the life-or-death choice of where to build their new home. This work has revealed how the process of evolution by natural selection, operating over mil- lions of years, has shaped the behavior of bees so that they coalesce into a single collective intelligence. This story about the bees also provides useful guidelines to human groups whose members share common interests and want to make good group decisions."

According to Seeley, as the title of his book suggests – *Bee Democracy* – bees make democratic decisions. He notices that when bees are preparing to swarm, they have the scout "compete" for the new swarm. More and more bees join in the dance behind the most appealing scout, until finally the one with the most dancers becomes the winner. This is how bees decide together where and when to swarm. Having read the book carefully and with great enjoyment, and having listened to Seeley's valuable lectures on the subject, I can say about him what he said about Lindauer, that he raises more questions than he answers.

In any case, I find it very characteristic and interesting how he explains the bees' decision-making process as a democratic decision, and however sympathetic his theory and convincing his evidence, based on what I have outlined above and my own experience, I am somewhat concerned that we are once again seeing a phenomenon characteristic of human society in our own time reflected in bee society. At the same time, even if what Seeley says is true, that bees make decisions collectively and that queen bees are not queens, but, as he writes

> ". . . a colony's queen is not the Royal Decider. Rather, she is the Royal Ovipositer."
> *(Thomas D. Seeley, Honeybee Democracy. Princeton University Press, 2011.)*

Do we see these wonderful creatures correctly, and when we look at them, do we see them, or rather ourselves?

There has been a total misunderstanding about bees for thousands of years, with them being seen as a society led by a king or queen, projecting human society onto bees rather than showing them as they really are – which, of course, is still the subject of research. The question is whether we can ever see nature as it is, or whether we always examine it through ourselves and our own definitions. It is as if humanity did not want to see nature when observing it, but rather to understand itself. It is no coincidence that I prefer to use Hungarian expressions in the art of beekeeping – mother for queen, bee family for bee society – because I find them both more accurate and less misleading. The central figure is much more of a mother than a queen, and the bee society is much more like a special family where everyone is actually everyone else's sibling.

Is it possible that the global technocratic, industrial, capitalist, and materialistic society that is currently emerging will once again want to see itself in the world of bees in some newer metaphor? That is, if there will still even be bees and pollinators at all, because, as we can see, the short-term interests and profits that are pursued above all else can easily wipe out bees and their habitat, and thus us as well.

But it may also be that self-reflexive observation, when the observer is aware that he himself is part of the observation and interacts with the observed, opening up new perspectives. Could it be that the global crisis will bring about local responses that create the possibility of survival? Perhaps our justified anger and despair at the escalating problems can be a new force that, when transformed, can give us the energy to find new answers to new challenges?

I don't know the answers to these questions, but I can pose them and also formulate them. And I also know that beekeepers live together with their bees, interact with them, and try to respond to each problem in different ways. Of the possible responses, the one that remains is usually the one that is useful, works, is energy- and cost-efficient, wise, and forward-looking, because just as bees do not like to do anything unnecessarily, neither do beekeepers like to work needlessly. The goal in both cases is survival, development, and prosperity.

However, I believe that we have always been very far from achieving mutual understanding between humans and bees, because humans constantly, consciously or unconsciously, anthropomorphize their environment, animals, and bees, thus endowing them with human characteristics that they do not possess. Bees have *bee characteristics,* even though they exist in the same world as us humans. We must observe, accept, and understand others in their entirety, and only then can we discover the similarities, intersections, and commonalities within our differences. For example, *they* are born, live, feel, desire, feed, develop, reproduce, communicate, struggle, and die, just like *us*, yet they do all this as *bees*, because they were born as *bees*. Both *bees* and *humans* have their place in the created world, and their interaction is so remarkable that I certainly see it with sincere admiration when I go out to the apiary and open a hive.

Bees and Sherlock Holmes, or the "Detective" Beekeeper

Beekeeping is an investigation. For the passionate beekeeper, inspecting each hive is like solving a detective story. When a good beekeeper opens a hive, he can tell from the smell, the buzzing, and what he can see what the bee colony is likely to be like. They say that a really good beekeeper doesn't even need to open the hive. Just by watching the bees flying in and out, he can see if there is any foraging; he listens and smells and knows whether or not to intervene. Unfortunately, I am not such an experienced beekeeper. I keep a diary, take notes, experiment, write down my most important and interesting observations, and from time to time I leaf through the pages to find out what the situation is. However, I admit that we need all our senses if we want to be really good beekeepers, especially if we strive to interfere as little as possible in the life of the bee colony. The truth is that it doesn't hurt to check on the bee colonies from time to time, say every ten days, to observe the signs that tell us exactly what is going on with them. Are they healthy or sick? Is there a queen or not? Is there open and capped brood? Is there honey and pollen? Do they need to be rotated? That is why I say that beekeeping is like detective work. We have to piece together the big picture from tiny clues – like in an essay or a Sherlock Holmes-style investigation – and figure out what is going on in the bees' lives, then to decide whether we need to intervene. In other words, we must adapt to the situation and always find the right solution. This is the superpower that proper beekeeping teaches us, which we could also call, using the term from modern psychology, *adaptive resilience*. I would like to note that sometimes we don't have to do anything, but are we capable of doing nothing? It is a very bad trait of humans to constantly want to "do" something, even though sometimes idle observation reveals more about the observed than intervention. Beekeeping is therefore an investigation, and every beekeeper is a little like Sherlock Holmes.

Sherlock and Watson in the apiary. In the title role: the outstanding actor Jeremy Brett, who suffered a tragic fate. Unfortunately, only a few photos remain from the film, as the recordings were destroyed under mysterious circumstances.

Nothing proves this better than the fact that when Sherlock Holmes, the greatest private detective in world literature, finally retired, according to the story, what else could he have taken up but beekeeping? Sir Conan Doyle, the creator of Sherlock, suffered from having created a character who later became a perfect "pop culture palimpsest figure," that is, one who came to life in thousands of forms in literature, theater, and film. It was particularly distressing for the author that most people believed that the detective really existed.

Anyone who is seriously involved in writing will eventually find that their writings take on a life of their own, especially in the case of short stories. Although I am primarily a poet and only secondarily a beekeeper, it is impossible to make a living from poetry, much less as a writer, and perhaps it is also pointless. I should note that it is difficult to make a living from anything else, but that is precisely why I learned to write short stories, plays, and essays like this one, and to give lectures to interested audiences on literature, life, and, more recently, the art of beekeeping. In the case of short stories, I have experienced how thin the line between reality and imagination can be. While theater can captivate the audience with the actors' performances, and catharsis can transform the viewer, it can still only distract the audience from their own reality, and when the curtain falls, the applause can easily dispel the beautiful dreams of the imaginary world. However, I have often found that people have taken the fictional story as a "reference to reality"; for example, they took a story I wrote about Nikola Tesla's years in Hungary and his laboratory in Pomáz as history. I should note that I myself have played on the interaction between reality and imagination, just as did Sir Conan Doyle and the master of us all, Edgar Allan Poe. It is true that when, after long philological work, I discovered, wrote about, and published the writings of Ferenc Békássy, people thought that I had invented the subject of my research and written all of his works myself, and they were disappointed when it turned out that our author had actually existed. It is often very difficult to decide what is a figment of our imagination and what is reality, sometimes even impossible, and it is better to focus on what a story is about, what message it conveys to us, and what we can learn from it. This is what happened with Sir Conan Doyle, who, after writing Holmes stories for many years, decided to retire his hero. And what else could Holmes have done but take up beekeeping?

Pendrills Cottage, East Sussex, England. This was the house where Sherlock Holmes retired and took up beekeeping between 1903 and 1917.

We have several sources for this. In the short story "The Adventure of the Second Stain" (1904), Watson says:

> "I had intended 'The Adventure of the Abbey Grange' to be the last of those exploits of my friend, Mr. Sherlock Holmes, which I should ever communicate to the public. This resolution of mine was not due to any lack of material, since I have notes of many hundreds of cases to which I have never alluded, nor was it caused by any waning interest on the part

of my readers in the singular personality and unique methods of this remarkable man. The real reason lay in the reluctance which Mr. Holmes has shown to the continued publication of his experiences. So long as he was in actual professional practice the records of his successes were of some practical value to him; but since he has definitely retired from London and betaken himself to study and bee-farming on the Sussex Downs, notoriety has become hateful to him, and he has peremptorily requested that his wishes in this matter should be strictly observed."

Memorial plaque on the wall of Pendrills Cottage in East Sussex, England.

But what is vital from this point of view is the short story *His Last Bow* (1917), which contains a very important scene after Watson and Sherlock drink Tokaj wine (!), and I quote:

"'But you have retired, Holmes. We heard of you as living the life of a hermit among your bees and your books in a small farm upon the South Downs.'

'Exactly, Watson. Here is the fruit of my leisured ease, the *magnum opus* of my latter years!' He picked up the volume from the table and read out the whole title, *Practical Handbook of Bee Culture, with Some Observations upon the Segregation of the Queen*. 'Alone I did it. Behold the fruit of pensive nights and laborious days when I watched the little working gangs as once I watched the criminal world of London.'"

Well, my investigation discovered in a local Australian newspaper that shortly afterwards, Sir Conan Doyle and his wife did indeed visit an apiary there, and he received letters from lonely ladies after he retired his hero, saying that they would like to be housekeepers in the detective's apiary. The story was continued in several later works, such as Laurie R. King's book *The Beekeeper's Apprentice*, or my personal favorite, the wonderful film *Mr. Holmes*, which we would definitely screen at the Bee Film Festival.

Sherlock Holmes The cover of a possible edition of the book The Practical Handbook of Bee Culture with Some Observations upon the Segregation of the Queen.

I myself began an investigation to find Sherlock Holmes' lost book on beekeeping, whose original English title is *The Practical Handbook of Bee Culture with Some Observations upon the Segregation of the Queen*. Perhaps it is no coincidence that I managed to find three different editions of this book, each with different content, all of which proved to be very interesting, and perhaps one day – if my other commitments allow – I will translate the most unusual of the three, which also has Hungarian references.

It is particularly interesting that in the legendary Sherlock film series – in which the tragically fated Jeremy Brett played the detective – scenes were shot in which Watson and Sherlock keep bees together, but for some strange reason these films were destroyed and only a few photos remain.

It is no wonder that Sherlock Holmes loved beekeeping, as he adored scientific problems, and there is no doubt that a beehive provides thousands of stories, from the life of bees and honey production to the study of the medicinal properties of various bee products. Sherlock loved solving seemingly mystical puzzles, and indeed, an apiary is full of mystical, intriguing cases, which tantalize us until we figure out when and, more importantly, why they happen. Of course, this does not detract from the fact that a real apiary is a genuinely mystical place. Sherlock loved to study the workings of complex criminal gangs, and there are few more complex "gangs" than a bee colony, especially many bee colonies together in an apiary. Finally, Holmes not only loved to investigate, but also to share, explore, expose, and even write about his discoveries in studies, thus spreading his knowledge to a wider audience. And we, the latter-day fans of Sherlock Holmes, love most when he presents how he came to the conclusion of what exactly happened and who the culprit was, even if we somehow suspected the latter from the very beginning. Because it is not just to know, we must also be able to explain what we know. After all this, we must try to apply our knowledge in practice, that is, to put it into action. Holmes's great achievement seems to have been beekeeping, as a Sherlock expert writes, after emphasizing that he practiced the virtues of patience, responsibility, preparedness, and perseverance in both his detective and beekeeper lives, he writes: "My opinion is that Holmes found the perfect pastime for his retired ages in keeping bees. He spent many years in almost constant danger, often under trying circumstances both for his body and his mind. For him observing the life of his bees and caring for them was genuine relaxation – something that he truly deserved after so many solved mysteries."

The creator of Sherlock Holmes, Sir Arthur Conan Doyle, and his wife visit an apiary in Australia on January 14, 1921.

The author in front of the Sherlock Holmes Museum on Baker Street, London, autumn 2025.

Why is it so important that Sherlock Holmes took up beekeeping after his retirement? Perhaps because, as I've mentioned, we are dealing here with a significant idol of modern society, a "palimpsest figure," that is, a constantly rewritten and idealized figure, or, if you will, a hero who uses science and common sense to understand mysteries, solve crimes, and, while he himself is in a state of constant turmoil, to restore order to a world in crisis. In our deranged modern society, the fictional detective has become a kind of holy madman who exposes crime and uncovers the truth. Thus, the beekeeping of Sherlock Holmes, this most famous detective, in his old age is not merely a coincidence. The place of beekeeping in society has always been – as I have tried to discuss several times in my essays on *The Art of Beekeeping* – a kind of mystical, or at least mystified, characteristic border guard or boundary role. As I wrote earlier, we find bee shamans in prehistoric cave paintings; we find in ancient Egypt that beekeepers had their own priestly caste and paid taxes only to the pharaoh; we read that Aristotle considered bees to be sacred creatures and beekeepers to be blessed and privileged; throughout the Middle Ages, bee society was idealized, and those who worked with bees were considered to be keepers of secrets. In Hungarian folk tradition, beekeepers were regarded as magicians, and even today, many people consider beekeepers to be strange but attractive madmen. Is it really crazy to be involved in beekeeping in the midst of a world in crisis? I don't know, but Sherlock Holmes' beekeeping shows that it is an activity connected with truth and thinking, contemplation and cognition, nature and science, in other words, very human things, which many people experience less and less in a world that is often deceitful and not very thoughtful.

A good beekeeper is therefore always a bit like Sherlock Holmes, and even if this seemed like poetic exaggeration at first, my readers can now see that it is true. The beekeeper observes, learns, thinks, investigates, discovers, and solves. He practices *adaptive resilience*, that is, the ability to adapt and be flexible in a given situation. Every bee colony is similar to others, yet each presents a unique problem to be solved, and the best problems are those that we can solve through mere observation, sensory perception, and often without intervention. A truly good beekeeper knows exactly when to intervene and when not to. In any case, it is precisely this constant observation that can lead us to view ourselves and our life with bees from both the outside and the inside. For example, this can also help us understand how bees and bee products heal us and others, but I will write about that elsewhere.

Bees and Visual Arts, or A Guide to Coexistence

1

In exploring the countless possible connections between bees, beekeeping, and the relationship between humans and bees altogether, we must focus on a somewhat arbitrary selection from the astonishingly rich collection that has accumulated throughout the history of art. This is especially true in the representation of bees in the visual arts, as this really takes us much further than a mere mechanical listing of works of art created on this theme from ancient times to the present. To select and become able to understand the message of these images, we must look a little "behind" them and glimpse the connections between works that are distant from each other in space and time.

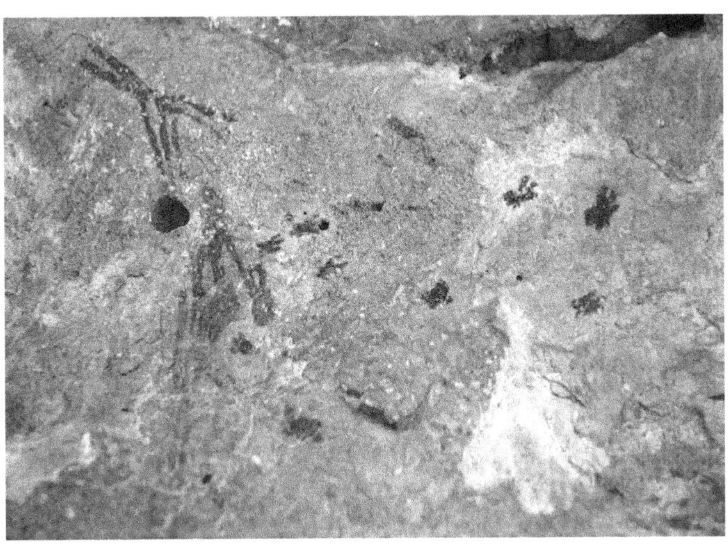

Honey hunter on a prehistoric cave painting discovered in the Cave of Spiders.

At the beginning of my series of essays, *The Art of Beekeeping*, I mentioned the *Cuevas de la Araña* cave, in the province of Valencia, Spain. This cave , better known as the *Cave of Spiders*, is filled with cave paintings dating back to the Neolithic period, the most ancient era of humanity. Here we find the first image depicting a human and bees together. Even though these cave paintings are estimated to be between ten thousand and eight thousand years old, we can clearly see a beekeeper, or rather a honey hunter, surrounded by bees. With a little imagination, we can even see the smoker (or basket) in his hand, not like the ones we use today, of course, but ones that serve the same purpose. I have written about all this and about ancient beekeepers before, but now that we are talking about the representation of bees in art, I must refer back to this ancient painting, and to another one as well. The other is located in the middle of the Sahara, in the caves of the Tassili n'Ajjer plateau in Algeria. The most daring estimate deems the mural to be ten thousand years old, while more conservative calculations hold that, along with many other rock paintings found in the area, it was created four to six thousand years ago. Scholars have distinguished at least three periods among the rock paintings and carvings, which tell the rich and tragic story of this region over thousands of years before our written history. Once upon a time, as difficult as it is to imagine now, the heart of the Sahara was a world full of water and life, with countless animals, plants, and with people who created these wonderful images. French researcher Henri Lhote discovered this collection of artifacts in the 20th century, and Jean-Domonique Lajoux was the first to photograph them and bring them to public attention. Today, the impressive prehistoric art of the Tassili Plateau is part of the world heritage, and deservedly so Among the images, we find a so-called bee shaman, a human figure with a bee's head, from whose body mushrooms grow – I wrote about this in my previously mentioned essay, noting that there are several such depictions on the Tassili Plateau.

Two bee shamans among prehistoric rock paintings found deep in the Sahara.

What is really interesting appears when we place the honey hunting cave painting discovered in the *Spiders' Cave* next to these bee shamans. In the former, humans and bees are separate, but in the other, the two become a single strange, human-animal chimera.

But what could the two have in common?

Perhaps that we can see both bees and humans in both?

Yes, although in one they are separate, in the other they are fused together. In one we can see a smoker, and in the other? Yes, mushrooms. Here may be the third, not so simple coincidence, because – as a Brazilian mycologist points out in his article (see:

Mariana Villani: Could the bee man shaman be more than we thought? An ethnomycology investigation) – the Tassili bee shaman is very often described, especially since Terence McKenna's seminal book (*The Food of the Gods*), as dancing with psychedelic mushrooms, and here these mushrooms are clearly visible around him. As is well known, shamans have used hallucinogenic mushrooms to achieve their visions since ancient times, and numerous studies on these mushrooms have been conducted in the fields of modern mycology and psychology. However, Mariana Villani points out that the bee mask is at least as prominent in the picture, so it is possible that these mushrooms are also related to bees. Is there a connection between bees and mushrooms?

I know beekeepers who collect bracket fungi (polypores) from old, fallen trees throughout the year and dry them to use in their smokers, claiming that it calms the bees better than plain smoke. I had no choice but to collect bracket fungi myself in the forests of the Pilis Mountains surrounding us, dry them, and try them as a fumigant. I admit that I didn't notice much difference compared to when I used the hollow insides of dead willow trees that had fallen along the Danube for smoking, although it undoubtedly gave the smoke an interesting aroma. Unfortunately, I didn't have any hallucinations, nor could I have from that mushroom, but it really was easier to work with the bees. Mycologists hypothesize that bees and fungi are more closely related than we think, and that the smoke from fungi affects them differently than it affects us. According to research by Paul Stamets and his colleagues (see: *Extract of Polypore Mushroom Mycelia Reduce Viruses in Honey Bees. Scientific Reports. Vol. 8. 1-6.*), polypore mushrooms and other mycelium networks that infiltrate hollow trees help bees strengthen their immune systems. A stronger bee immune system is better able to resist even dreaded diseases which can lead to the collapse of bee colonies. such as the deformed wing virus spread by Varroa destructor mites. Perhaps this is why bees have been observed to sometimes collect not only from flowers but also from fungi, and why they have always favored the insides of hollow, older trees, which were not only their natural home but also that of the fungi with which they have coexisted for millions of years. So, when we put bees in plastic hives, we deprive them of beneficial fungi, while harmful spore-forming single-celled organisms – such as the dreaded Nosema – come uninvited.

We find numerous separate depictions of bees in the cave paintings of the Sahara, but in any case, the images of bee shamans are the strongest evidence of the early coexistence of bees and humans, or – to be more precise – borrowing and extending a political-philosophical concept to our main point: their coexistence. This fact of coexistence appears like a connecting golden thread when we examine the relationship between bees and humans in the visual arts of different eras, often only at the level of symbols, allegories, but, more recently, in very specific references. So, let's take a closer look at how this appears in the visual arts of later periods, such as the Middle Ages and the present day.

2

I wrote earlier about how the bee hieroglyph was used in the ancient Egyptian empire, but we could also consider one of the seven wonders of the ancient world, the Temple of Artemis in Ephesus, which housed a statue of the ancient goddess Artemis. This unique statue was a true mother goddess statue, an interpretation of Artemis completely different from what we know from our studies of mythology today. According to myth, she nourished life in the world from her countless breasts, and ancient depictions of bees appear on her clothing. Among the ruins of the ancient city of Ephesus, we find reliefs in the shape of honeycombs, all of which indicate that the later cult of the goddess Artemis, built on the local mother goddess cult, preserved its close connection with bees. Let us remember that for a long time, honey was the only sweetener, and bees were seen as sacred and pure creatures, created in Egypt by the sun god Ra, and in ancient Greece by Zeus from the body of a dead bull, as a sign of forgiveness to Aristaeus, who thus became the first beekeeper after capturing the bees. Of course, this is "only" mythology, but it shows very well how important bees were in ancient times and how much they were honored. Honey was also significant in the Old Testament, and this was passed on to Christianity through various stories about Jesus, as I have already written about.

However, art in the Middle Ages and the Renaissance revived and revitalized ancient stories and myths, filling them with new content, while also turning toward tangible everyday life, sometimes in an extremely practical way.

It is worth looking at two images from a similar period, one is Lucas Cranach's painting *Cupid Complaining to Venus*, created in 1525, and the other is Pieter Brueghel the Elder's drawing *The Beekeepers*, from 1568. Just like the honey hunter found in the Spider Cave' rock painting and the Saharan bee shaman rock painting, it is worth placing these two side by side and looking closely at both works of art, because we can observe something quite extraordinary here.

Lucas Cranach the Elder: Cupid Complaining to Venus. 1525.

Cranach's painting *Cupid Complaining to Venus* was inspired by a poem by Theocritus recounting an ancient myth. Theocritus was an early bucolic poet who had moved to Alexandria, one of the world's first cities with a population of over a million. He longed to return to the countryside, to his lost childhood, to the lost Arcadia, where he was born and raised. From this longing, he created pastoral poetry, dialogic verse, and perhaps even fictional literature itself. Virgil took the idea of eclogues from Theocritus, and the excellent

Hungarian poet Miklós Radnóti took it from Vergil. When did Miklós Radnóti write eclogues? During World War II, when he was taken to a labor camp because of his Jewish origins, then to a labor camp in Bor (Serbia), then, finally he was killed in Abda after a forced march. It is precisely this Bor notebook that shows that there were some who managed to remain human in the face of inhumanity. And it is in this notebook that we find Radnóti's most beautiful and moving poems, such as *Forced March*, in which the poet writes:

"(...)Oh, if I could believe that I haven't carried what is to be cherished

only in my heart; that in the real world there is a return, a home;

tell me it's all still there: the old cool verandah, the bees

of peace, of stillness humming, while the plum jam cools (...)"

"The bee of peace..." writes the poet, and most of us probably do associate this image of bees with peace, security, abundance, and have done so since ancient times. This is related to the fact that honey produced by bees was one of our first natural sweeteners, and that humans instinctively associate sweet tastes with positive things, since breast milk is also sweet. Humans therefore instinctively associate sweet taste with security, maternal love, and easily and quickly absorbed energy, which also poses a huge danger and health risk, especially with the advent of industrial sugar production and high sugar consumption. I would like to note that honey and industrial sugars are metabolized differently in our bodies, producing different effects, but of course, like everything else, both sugar and honey can be harmful if consumed in excess. I will write about the history of the great war between honey and sugar elsewhere.

Returning to Cranach's painting, it is clear that this is about something else entirely. In the painting, of which there are several versions, we see Venus-Aphrodite, the goddess of love, as a beautiful nude, clinging to an apple tree – just like Eve at the moment of the Fall – while next to her, little Cupid, surrounded by buzzing bees, grabs a honeycomb with a bitter expression on his face after having been stung, and visibly complaining, as the title of the painting suggests. As I mentioned, the painting was inspired by a poem by the ancient Greek poet Theocritus. This poem is better known in a 19th century translation and an idyll titled *The Honey Thief*:

> "A cruel bee once stung the thief Eros as he was plundering honey from the hives, and it made all his fingertips tingle. In pain, he blew on his hand, stamped on the ground, and jumped about, and he showed his wound to Aphrodite, complaining that such a small creature as a bee should inflict such wounds. His mother laughed and said, "Aren't you just like the bees, so small and yet so wounding?"
> *(Theocritus. IDYLL XIX. Κηριοκλέπτης. The Honey Thief.)*

Cranach's painting attempts to interpret this verse, but also to expand its possible interpretations. For example, if we know that syphilis was rampant in Europe when the painting was created, we can easily interpret it as the painter's attempt to draw attention to the fact that the small pleasures that physical love can bring can very easily lead to great suffering.

Pieter Brueghel the Elder: The Beekeepers. 1568. Ink on paper (Staatliche Museum, Berlin).

There are no roses without thorns, just as there are no bees without stingers.

At least here in Europe, because elsewhere there are bees without stingers. It is characteristic and typical that Cranach depicted the sweetness of love and the suffering that follows in its wake with honeycomb and bees, as the images of honey and bees were very often associated with physical pleasures in poetry as well as in painting. The didactic and allegorical intent of the painting is unquestionable, and rather than analyzing it further, I would like to remind you that this painting was created during the Reformation, in the workshop of a painter who later became Martin Luther‘s exclusive portraitist. In Cranach‘s painting, the female nude looking seductively at us from among the bees stands in stark contrast to Brueghel‘s drawing, created just a few decades later, in which the beekeepers are fully clothed and even cover their faces with basketweave woven into their hoods while they work with their hives. This drawing fits into Brueghel's series depicting rural life. Behind the beekeepers, we see a man clinging to a tree and plundering birds' nests, with a contemporary city in the background.

Yet there is something peculiar about this drawing, something that has changed completely since Cranach's painting. While there, the bees were characters in a mythical, allegory, here they are very concrete, part of everyday life. Of course, Brueghel also has paintings in which bees play a mythical role, such as *The Fight Between Carnival and Lent* from 1559, in which we see a beehive on the head of Lent, symbolizing diligence and self-restraint. But returning to the drawing *The Beekeepers*, one cannot fail to notice the dynamism of the image, in contrast to the static nature of Cranach's painting, and while *The Beekeepers* painting is quite "folk art" and depicts contemporary beekeeping with its tools and the hive, *Cupid Complaining to Venus* is based more on classical culture, and although the secret protagonists of both paintings are bees, each approaches them in a completely different way. If you like, Cranach's painting shows "holy" or "teaching" bees, while Brueghel's drawing shows the "profane" side of everyday beekeeping. In one, the bees are holy teachers who can impart wisdom with a single sting from their tiny stingers, while in the other, they are domesticated animals, from whose stingers resourceful beekeepers protect themselves with this strange attire. Let's look at the two images again and ponder how the relationship between humans and bees, humans and nature, has changed.

What role do bees and nature play in our lives? Is it sacred or profane? Do we learn from them or exploit them? Which image is closer to us, which is further away? Which do we like better? Which do we like less? Could it be that the one that seems further away, the one we like less, is actually the one we need to understand better?

However, nowhere else are painting and beekeeping so closely linked as in Slovenia, where for centuries the gates of beehives have been decorated with special paintings, and the most beautiful collection of these paintings can be found in Radovljica, in the beekeeping museum. That's why we had no choice but to visit this slightly remote but extremely beautiful Slovenian town, located north of Ljubljana and south of Mount Triglav.

3

Folk art and beekeeping have never been so beautifully combined as on the front panels of traditional Slovenian container hives. Slovenia is one of the Canaans of beekeeping, with a wonderful beekeeping tradition. It is the home of the Carniolan bee and Anton Janša (1734–1773), Queen Maria Theresa's chief beekeeper, who started beekeeping education in the empire. His teachings were followed by, among others, our own Sámuel Tessedik, who was the first to talk about how to keep bees and harvest honey without killing the bees, because until then it was common practice to simply destroy the bee colonies from which the honey was taken. Anton Janša also used such Slovenian-style small hives, which were the characteristic beehives of the Slovenian countryside, and which can still be seen today when visiting our neighboring country. The entrances to the many small hives stacked on top of each other had to be marked in some way, on the one hand so that the bees would not move from one to the other, and on the other hand so that beekeepers could distinguish between them. For this purpose, paintings were made on simple wooden boards, mainly depicting various religious stories, but there are also many secular images.

Typical Slovenian beehive paintings.

The first such beehive paintings date back to 1758 – although there were probably earlier ones – and depict the Virgin Mary and the baby Jesus among flowers, wearing a crown. The painted beehives have a characteristic, what we might call "peasant Baroque" style, and their naive paintings tell a wide variety of stories. A huge collection of them can be admired in the beekeeping museum in Radovljica. The motifs are mainly divided into two groups: sacred Christian stories and profane images taken from rural life. The former mainly feature biblical stories, while the latter often include humorous and even surreal images. You can learn more about them in Petra Bole's book (Petra Bole: *Living Together. About Bees and Mankind*. Radovljica. 2021). This beautiful tradition is still practiced in Slovenia today, and working with bees is motivated not only by honey production and profit, but also by spiritual health. Slovenia is not only a stronghold of beekeeping, but also of healing with bees and bee products, or apitherapy.

Hundreds of years of uninterrupted beekeeping, time spent in apiaries, and developing folk medicines has helped Slovenians realize that working with bees calms the soul. They have noticed not only the health benefits of bee products such as honey, propolis, and pollen, but also – perhaps while painting these hives – that bees are good for mental health. They studied and practiced bee therapy, including bee venom therapy, with particular reference to Philipp Terč (1844-1917), a doctor and beekeeper of Czech origin who settled and practiced in Maribor. From him, our own Bódog Félix Beck also learned, and this was the same Bódog Félix Beck who first penned the term apitherapy and practiced bee venom therapy in America. At the dawn of the modern era, folk traditions shed light on knowledge that seemed to have been completely lost: the healing power of bees and bee products for maintaining health and curing illnesses. All this, however, takes us far from our topic, yet it clearly shows how closely art and life are intertwined, and how much the role of bees has been reevaluated by the time we reach contemporary art. This is especially true because by the end of the 20th century, not only honeybees but all pollinators have been seriously endangered by excessive human intervention, globalization, and chemical-based industrial agriculture, and in our century, this is further exacerbated by the unpredictable consequences of climate change.

In today's art world, there are many artists who deal intimately with bees and nature, such as Tomáš Gabzdil Libertíny (born 1979). The Slovak-born artist creates beeswax sculptures, mainly vases, which he completes in collaboration with bees. The artist describes the process as "slow creation," which stands in stark contrast to fast and mass plastic and serial production, while also drawing attention to the importance of beeswax.

American painter Lea Bradovich (1955), whose art I have been following for a long time, combines the world of Renaissance painters with surreal visions. Lea Bradovich's paintings are unique portraits depicting people together with bees, often completely fused together, as if they were distant descendants of the Saharan bee shaman. Observing Lea Bradovich's work led me to the astonishingly beautiful collection of essays by contemporary American essayist and poet Heather Swan, in which she writes about the paintings of Lea Bradovich, who is of Bosnian origin:

> "What drew me to these images most forcefully was their complex ambiguity. One might read these paintings as images of people using bees as decoration, which certainly seems to reflect our current use of the bee as a tool or a symbol divorced from its ontology. But another way to see them—indeed, the way I think Bradovich intends us to see them—is as images of humans honoring the bee, perhaps representing something like a bee renaissance. The bees are quietly working the comb in several of the images, the human and the bee both seemingly at peace. What I choose to see is coexistence."
> *(Heather Swan, Where Honeybees Thrive: Stories from the Field (Animalibus: Of Animals and Cultures). University Park, Pennsylvania: The Pennsylvania State University Press, 2017.)*

In this special book, Heather Swan also writes about several contemporary artists whose work relates to bees and about her own experiences with beekeeping. She concludes by saying that:

> "A movement toward coexistence is what I have encountered over and over on this journey of learning about bees and the people who love them. Nonetheless, I recently read an article confirming another 37 million bees dead in Canada. The article was careful to avoid drawing definite conclusions as to what killed them, but the beekeeper quoted in the piece is convinced that neonicotinoids are the culprit. Maybe the bees won't win this fight. Maybe the electronic bee will be

> a godsend. But many artists, scientists, beekeepers, farmers, activists, and citizens are creatively advocating for the bees; a few of them are featured in this book. In a healthy beehive, labor is divided among the bees, some gathering nectar and pollen, some nursing new bees, and some building comb as others produce honey and wax. To change the way we live on this planet will take everyone doing what they are best at doing. Writing this book has provided scaffolding for my belief in the human potential for creativity, connection, compassion, wonder, cooperation, joy, and action. Is it possible that we are on the verge of an environmental renaissance? Time will tell."

Time, I might add cautiously, of which neither bees nor humanity have much left.

In connection with all this, I must mention one last painting, which is Hungarian painter Márta Kiss's *Beekeeper*. The painting depicts a swarm being caught, with the beekeeper placing the bees from the bag into the swarm box, their flight indicated by delicate embroidery on the painting. Although the painting is seemingly related to Brueghel's beekeepers, the purple background elevates the painting to a much more metaphorical, dreamlike plane, speaking simultaneously of the domestication of "wild" nature and of some inner feeling, some kind of vision that is both unsettling and beautiful, and which is the unique world and experience of the 21st century. This *Beekeeper* is an inner image that has become external, both beautiful and disturbing, perfectly expressing the uncertainty in which beekeeping will exist today, which is why it is incredibly important, because it points out that we are at a crossroads, as we have been so many times in human history, from prehistoric cave paintings to the present day. The wonderful works in the history of visual arts, at least to me, seem to point to these crossroads both instinctively and consciously.

The main message of *The Art of Beekeeping* is the possibility and recreation of the coexistence of bees and humans, nature and humanity, both on an instinctive and conscious level, because understanding and experiencing this is the key not only to the survival of humanity, but also to the survival of the world, if we want to overcome our own limitations and destructive tendencies.

Márta Kiss: Beekeeper. 34 x 24 cm. Cardboard, ink, tempera, colored pencil, embroidery thread. 2022.

The Healing Power of Bees, or the Meaning and Purpose of Bees and Bee Products

1

For two years, I wrote essays for *The Art of Beekeeping*, which aimed to highlight apiculture, or the cultural relationship between bees and humans from the beginning to the present day. My essays have been published, sometimes in several installments, in the monthly journal *Beekeeping*, and slowly the separate stories came together to form a single book. All the while, I continued to keep bees myself, slowly and consciously developing our small family apiary and my own beekeeping skills. Every spring, I spent an increasing amount of time in the apiary, reading more and more specialist literature and – something that is priceless – getting to know excellent beekeepers in person, from whom I learned a great deal. As I neared the end of writing *The Art of Beekeeping*, there were still many questions left unanswered and stories left incomplete. Many of these I had noted down at the beginning of my research as topics to be researched, explored, and written about. Two of them are the most important, and I would definitely like to discuss them, albeit briefly: one is the healing power of bees, and the other is the "purpose" of bees. One is a practical question, the other more philosophical, but they are related. Both questions are extremely complex and will likely lead to further research questions, so I will probably conclude this series of essays not by drawing conclusions, but by pointing to new paths and other possibilities. After all, my previous writings have also had the goal of broadening intellectual horizons, reintroducing traditional or forgotten ideas, insights, and stories, as well as presenting new ones.

When you go out to the apiary, you are filled with unusual, pleasant, and positive feelings. A sweet scent wafts from the hives, and the air is filled with a golden glow and a dense buzzing. If you are even slightly sensitive, you will begin to feel that the proximity of bees has healing powers. Humanity has been using this healing power since the beginning of time, and today it is called apitherapy.

When talking about the healing power of bees, we must first clarify its complex concept. Healing with bees and bee products is called apitherapy. The combination of the Latin word "apis" and the Greek word "therapy" was first described by the Hungarian physician Bódog Félix Beck, who later became very important in the field of apitherapy, especially regarding the much researched and controversial bee venom therapy. I wrote earlier about bees and mental health, that people knew from the beginning that bees might have something to do with maintaining human health. Perhaps this practical knowledge was based on the belief that bees were of divine origin, were "blessed," the tears of the sun god Ra, or the "daughters of the Sun". Bees were even considered examples to be followed in a social sense. Perhaps it was the realization that, in addition to the healing substances they provide – for example, honey – bees are indispensable in agriculture through pollination, i.e., in providing us with food, and this too led to their popularity and esteem. This is because bee products such as honey, propolis, beeswax, pollen, and even beehive air, not to mention the larvae themselves and, in particular, bee venom, all have an undeniable and proven positive physiological effect on humans. Their extraction, effects, and applications are now understood under the expanding concept of apitherapy.

So, it is worth taking a look at the history of healing with bees, highlighting just a few important details to see that apitherapy did not begin today. Honey was already of great importance in prehistoric times (see the essay entitled *The Bees of Our Ancestors*), and in ancient times Hippocrates (460-377 BC) also used honey and bee venom for healing. He can be considered the first source of external and internal treatment with honey, as we can read in the following book on natural remedies:

> "Honey was a valued and frequently used natural substance by representatives of Hindu, Greek, and Arab medical schools. Indian medical books devote several pages to its healing properties. According to popular belief, Goliath and Hercules consumed honey daily as part of their diet. (...) Among Hippocrates' most important remedies were honey and water, as well as a mixture of honey and vinegar. For centuries, honey was a typical carrier substance for medicinal preparations, and it was a simple way to preserve plant extracts. Due to its mineral and vitamin content, it is still an important restorative tonic today."
> *(Dr. Judit Szarvasházi, Man and Nature in Medicine. The Use of Animal and Herbal Materials from Antiquity to the 20th Century. 2007.)*

Pliny the Elder (24-79 AD) discussed the healing properties of bees in his *Naturalis Historia*, while Galen (129-199 AD) used bee venom in several different medicines. The Frankish emperor Charlemagne (742-814), and Ivan the Terrible (1530-1584) were both treated with bee stings, and we know that French doctors were already collecting bee venom in the seventeenth century.

In 1859, a French doctor named Desjardins published the first scientific paper on bee venom therapy for the treatment of rheumatic complaints, but the breakthrough came only with the Czech-born beekeeper and doctor Dr. Filip Terč (1844-1917), who, after moving to Maribor, Slovenia, a place with a long tradition of beekeeping, treated his patients with bee stings and published his findings in 1888 in a paper entitled "Report on the Special

Relationship between Bee Stings and Rheumatism." During the same period, research was also being conducted in Russia and Austria, but bee venom was not isolated until 1928. In 1935, Bódog Félix Beck published a detailed account of bee venom therapy, which he was already practicing in New York at the time, reportedly with success. His book, *The Bible of Bee Venom Therapy*, of which I have the 1997 edition, is incredibly detailed and exciting.

BEE VENOM THERAPY

Bee Venom, Its Nature, and Its Effect on Arthritic and Rheumatoid Conditions

BY
BODOG F. BECK, M.D.

There proceedeth from their bellies a liquor wherein is a medicine for men.
THE KORAN, xvi, 71.

D. APPLETON-CENTURY COMPANY
INCORPORATED
NEW YORK LONDON

Bódog Félix Beck: Cover of the first edition of his book, The Bible of Bee Venom Therapy, published in 1935.

Bódog Félix Beck was a Hungarian Jew, who was born in Baja in 1868 and died in New York in 1942. He continued Terč's work and, as far as we know, was the first to establish the term apitherapy and bring the idea of healing with bees and bee products to the United

States. It is clear that, in the modern era, apitherapy was primarily understood to mean the medicinal use of bee venom. This naturally limits its possibilities, but I will return to this topic later. I should note, however, that Beck wrote an extremely valuable book on honey entitled *Honey and Health*, which was published in 1938 but was unfortunately falsified after Beck's death. The latter edition is better known and more accessible, and János Körmendy-Rácz also wrote about this and about the life of Bódog Félix Beck in his book *The Life of Dr. Bódog F. Beck. Lilli. Nagykovácsi. 2019).*

Bódog Félix Beck treats one of his patients with bee stings in his New York clinic office. 1938.

The most important aspect of apitherapy, as with all healing and remedies, is the principle of "do no harm." Modern humans, who have become estranged from nature, are unfortunately prone to various allergies, allergic reactions, autoimmune diseases, and diabetes. Although many of these conditions can be treated with apitherapy, they can also become significant risk factors.

Bee venom therapy has been controversial from the outset, as it is not without risk, since it can cause allergic reactions in some people, and there have been other problems with it, as Bódog Félix Beck discovered during its early applications. In bee venom therapy, it was found that different effects are achieved when bees are used to sting so-called trigger points, or what are now called apipuncture points, in the "traditional" way, as opposed to extracting the bee venom with great difficulty, putting it into an injection, and administering it to patients. Experience has shown that traditional bee stings are more effective, which foreshadowed the main problem, namely that although various bee venom-containing preparations, such as ointments, can be produced and undoubtedly have an effect, nothing compares to stinging the appropriate points with live bees. The pharmaceutical industry has largely lost sight of bee venom therapy. Moreover, when I tried to follow up on whether we could continue such experiments at home, i.e., whether we could investigate the medicinal use of bee venom, the first and most important problem was that it is difficult to synthesize, extract, and license such an allergenic substance. This is an extremely difficult task because apitoxin, or bee venom, is a general cell toxin. Accordingly, it was difficult to imagine that it could be used to make a specific medicine. Its production by peptide synthesis is not such a big challenge, but because it is a strong toxin, no one in Hungary would take it on. But even if someone did, it would certainly require strict licensing. In other words, it would be difficult for pharmaceutical companies to produce an effective, licensed, easily manufactured, and, above all, *marketable* drug from bee venom. Admittedly, all I wanted to achieve was to experiment with it on patients with rheumatism, multiple sclerosis, and autoimmune diseases. Since I was able to start on this, I examined the house with beehive air in cases of asthma, sleep disorders, and anxiety and collected data on the results. Admittedly, this is not as dangerous as bee venom therapy. Why would I want to experiment with bees? Because I wanted to find out if they really healed and what their ultimate goal was.

office. 1938.

2

Looking beyond our small country, we can see that bee venom therapy is now used in many places around the world, primarily for patients suffering from rheumatism, multiple sclerosis, and autoimmune diseases. This has probably been the most controversial area of apitherapy from the outset, for many reasons, but mainly because it cannot be replaced by a specific drug in traditional interventions, because it is not risk-free, and because – like all bee products – its therapeutic effects are not immediate, but develop over time and therefore take place through regular use and treatment. This is why apitherapy is a good word. The use of bee venom therapy requires sufficient expertise, prior allergy screening, preparation, and, if possible, medical supervision. Modern people are accustomed – mainly thanks to painkillers – to taking a pill or getting an injection and achieving immediate results, while the vast majority of natural remedies only exert their healing power through regular, long-term use. Perhaps the increasingly widespread forms of psychotherapy are making people realize that they need to think in terms of processes and connections when it comes to both physical and mental health. Just as a small spoonful of honey is wonderful, delicious, and full of energy, the healing effects of honey can only be experienced with regular - and moderate - consumption, as is true for propolis, pollen, hive air, and bee venom. Furthermore, as some apitherapy experts have rightly pointed out, bee products are not miracle cures, they are not good for everything, and even for what they are good for, they may not bring about a complete cure, but often only some improvement. Recommended reading on this topic is István Gillich's recently republished book *Apitherapy*. Just like bees, humans are complex creatures. The interaction between bees and humans is so complex and inspiring that even as a poet, I find it a great challenge to write about it in these essays. But let's ask the question again: do bees really heal, and is that their "ultimate goal"?

The most common criticism, of course, is whether there is any scientific basis for apitherapy, i.e., healing and therapeutic effects using bees. Science is the religion of modern man. People believe science more than their own experiences, but we must give credit to science for not being a fixed dogma but rather a constantly evolving dialogue that can correct its own mistakes. Therefore, science sometimes provides a broader perspective than religious dogma, not to mention that, in the best case, it is based on repeatable experiments, research, and the conclusions drawn from them. Here we must go a little further into the history of apitherapy and repeat that when Bódog Félix Beck wrote his exhaustive and very detailed book on the healing properties of honey and the harmful effects of sugar, which was already being produced industrially and consumed in large quantities at the time, he had previously written another book about the bee venom therapy he used. Unfortunately, this extraordinary Hungarian-born physician and author is less well known in Hungary, perhaps because Beck spent most of his life in New York, where he practiced medicine. He lived next to Central

Park and kept bees to be used in his therapies in a hive in his window. One of Beck's students was the excellent Vermont beekeeper Charles Mraz, who contributed greatly to the popularization of apitherapy in the United States. They performed many apitherapy treatments and taught members of the American Apitherapy Society (AAS), which he inspired. In his book *(Health and the Honeybee. Honeybee Health Products, Vermont. 1995)*, he refers several times to Beck's bee venom therapy and to Beck himself. Charles Mraz obtained queen bees from Buckfast Abbey in England, as we have learned from recollections, and he also delivered the new bees to Beck in New York. Mraz himself attempted to conduct clinical trials with bee venom therapy, but for reasons similar to those in our country, these trials were discontinued. While Beck approached apitherapy with extraordinary erudition and a medical attitude, using primarily honey and bee venom, Mraz, as a beekeeper, used bees for healing in a practical way, following Beck's guidance and based on his own experience, with what we might call "good ol' common sense." Of course, Beck and Mraz could be called biased, but the comprehensive and very accurate scientific paper published in 2022, which summarizes and lists bee products and their medicinal effects in detail, referring to experiments conducted in different parts of the world and summarizing knowledge and experience, cannot be called biased in the least. The title of the defining paper on 21st-century apitherapy is *An Overview of Apitherapy and Its Clinical Applications*, in which the authors write:

> "Humanity has benefited from bee products over the centuries for treating and preventing various illnesses, and apitherapy has been employed in several countries as complementary medicine. This review aimed to discuss scientific research and clinical trials using bee products and their relationship with apitherapy. (...) The most common applications of bee products in apitherapy are presented, as

well as cases of allergy to bee products and apitherapy for treating allergies. Standardization of bee products and their use in research and apitherapy are discussed.

"Conclusion: Apitherapy is practiced in some parts of the world, bringing benefits for healthy individuals and patients, with no clear consensus on its application according to the world regions or a prevalent use of bee product and treated disease. [*note from author: the frequency of occurrence of the condition affecting the given population]* Different recommendations regarding the use of bee products are found and people ingest different amounts of bee products once or several times a day. Although we have advanced a lot about the knowledge of bee products, it is imperative to exploit their potential and standardize their use, communicating the results in scientific and alternative events to reinforce the exchange of information between beekeepers, researchers, apitherapists, nutritionists, physicians, sellers and consumers of bee products. If not, we will always be working separately, without complementing our expertise."
(Wilson Antonio Weis, Nicolas Ripari, Fernanda Lopes Conte, Mariana da Silva Honorio, Arthur Alves Sartori, Ramon Hage Matucci, Jos´e Maurício Sforcin: An overview of apitherapy and its clinical applications. São Paulo State University (UNESP), Institute of Biosciences, Department of Chemical and Biological Sciences, Campus Botucatu, 18618-691, Brazil 2022. Published in Phytomedicine Plus 2. (2022) 100239.)

The paper then reviews and summarizes in detail all assessable scientific research to date on the medicinal effects of bee products, with particular emphasis on honey, propolis, and bee venom. It is worth mentioning the interesting and inspiring ideas on apitherapy of our contemporaries Stefan Stângaciu, who is of Romanian origin but carries out invaluable work in Germany, and the papers by Catherine Ballot-Flurin from France. Mihály Hevesi also does valuable work in spreading knowledge on this subject in Hungary. However, all of this goes well beyond the scope of *The Art of Beekeeping* and points in a completely different direction, which I may write about in more detail later, but it clearly shows that bees play a key role not only in pollination and agriculture, but also in healing and maintaining our health.

After this great adventure from prehistoric times to the present day, we must ask the poetic question: what is the "purpose" of bees? Why did God create them? Or why did nature create them? This seemingly naive question was asked long before me by the great writer Leo Tolstoy, who was also a passionate beekeeper and a great lover of nature and creation. Tolstoy writes this in one of the unforgettable pages of *War and Peace*:

> "A bee settling on a flower has stung a child. And the child is afraid of bees and declares that bees exist to sting people. A poet admires the bee sucking from the chalice of a flower and says it exists to suck the fragrance of flowers. A beekeeper, seeing the bee collect pollen from flowers and carry it to the hive, says that it exists to gather honey. Another beekeeper who has studied the life of the hive more closely says that the bee gathers pollen dust to feed the young bees and rear a queen, and that it exists to perpetuate its race. A botanist notices that the bee flying with the pollen of a male flower to a pistil fertilizes the latter, and sees in this the purpose of the bee's existence. Another, observing the migration of plants, notices that the bee helps in this work, and may say that in this lies the purpose of the bee. But the ultimate purpose of the bee is not exhausted by the first, the second, or any of the processes the human mind can discern. The higher the human intellect rises in the discovery of these purposes, the more obvious it becomes, that the ultimate purpose is

beyond our comprehension. All that is accessible to man is the relation of the life of the bee to other manifestations of life. And so it is with the purpose of historic characters and nations."

(Leo Tolstoy: War and Peace)

The author of the series of essays on the art of beekeeping in his own apiary. Photo: Bence Illyés.

There is no doubt that the ultimate goal of bees is as inaccessible as our own ultimate goal. Nevertheless, I dare say that the health-preserving and healing effects of bees and bee products are infinitely valuable and essential for the survival of our world, indeed of humanity and of individual human beings. But that is another story.

Bees and the poet

"Marcus Aurelius says that by the age of forty, a man with a spark of intelligence has experienced everything and knows everything that has happened to people in the past and what may happen in the future."

This is what Sándor Márai writes in his book *Füveskönyv (The Grass Book)*, and it is worth reflecting on for those who are not yet forty, those who are over forty, but especially those who are just turning forty. I believed this a little in May 2021, when I started my first year of production with seventeen bee colonies, more as a poet than as a beekeeper.

In 2015, my wife and I moved to Szentendre, to Kada Peak, from downtown Budapest, where we had lived for seven years. Before that, I attended the University of Szeged. According to my degree, I am a *philologist*, that is, a "literary gentleman", one who tinkers with texts, as Antal Szerb called his peers in *The Pendragon Legend*. Incidentally, although I was born in Eger, I grew up in a village in the Little Hungarian Plain. My parents had a large garden, and my brother and I did a lot of gardening. I still have a close and loving relationship with my brother, and he has helped and supported me a lot in beekeeping. I consider myself lucky because, despite all the difficulties, I have been blessed to have such a good brother, a few excellent friends, and a wise wife. My mother sent me to a secondary school specializing in food industry (dairy), even though I was already writing poems at the age of ten and had no intention of becoming a farmer or cheese maker. In the end, however, through continuous learning, writing, reading, gardening, and beekeeping, food production and manufacturing returned to my life.

Weiner Sennyey Tibor with his wife (Virág) and his dog.
The photo was taken by István Bielik

Around 2019, while walking my dog, I saw Uncle Misi (Mihály Hudek) bustling about next to some beehives on the other side of the hill. He later became my mentor. Then, at the end of September 2020, my gardener friend noticed two abandoned bee colonies on a dilapidated farm hidden in the outskirts of one of the surrounding villages. After finding out that the farm had been abandoned for years and was overgrown with weeds, we brought the bees home and tried to prepare them for winter from scratch, which, as every beekeeper knows, is very late. One colony was from Krajina, and my wife named the queen "Etelka," while the other was Italian, and she named their queen "Julia." We kept them in a closed space – a cellar – during the winter, and their chances of surviving were very slim. In the end, Etelka's family survived the winter without any problems and became a great productive family, but unfortunately Julia gave up after the first warm days followed by a cold spell, so right at the beginning I experienced both success and failure in beekeeping. Bees are great teachers, and I don't think I'm the only one who thinks so.

Originally, I wanted to take it step by step, expand slowly, and I imagined beekeeping as a kind of hobby that I would learn in my old age, but things turned out differently. From the very beginning, my connection with bees was not a means of livelihood, but a special passion, just like gardening. And I found I was interested in beekeeping for three other reasons.

First of all, because I have always loved learning, and I felt that I could acquire a lot of new knowledge as an adult. This is undoubtedly true.

Secondly, because I see, know and experience the ecological crisis that affects us all. My first book of poetry, published in 2005, was titled *Rethinking Our Relationship with Nature*, and as this title suggests, I have long been concerned with how we can respond to the challenges of the 21st century, not only on a community level, but also on an individual level. I now know that, on an individual level, everyone should live an ethical, moderate life, which is greatly helped by some kind of spiritual path, such as meditation or prayer, but faith is a private matter, whereas an ethical life also has an impact on our environment. We can make it better. We have to start the change in our own little lives if we want to have a livable world, in which, for instance, neither pollinators nor we die out. This basic stance stems from my love of my country – I consider Hungary and the whole world to be beautiful, lovable, full of wonders and knowledge, and worth protecting. And I think this way because I love life, animals, plants, and people. Nowadays, it has become fashionable – unfortunately, partly with good reason – to be misanthropic and to blame and hate humanity and people for everything. I do not share this opinion, because I, we, and our loved ones are part of humanity and the created world. Humanity, and especially individual human beings, have accomplished many wonderful things alongside the many atrocities. Science, the arts and crafts, that is, the quiet history of peace and prosperity and its heroes, are somehow not so well known. Perhaps because the *Book of Life*, in which the names of the righteous are recorded, is not on this earth. I have always been more interested in beautiful works of art and artists, useful scientific discoveries and scientists, good work and hard-working craftsmen than in the stories of murderers and villains, or, as Jorge Luis Borges writes, “the universal history of infamy”. I was more interested in the world of hard-working, talented, humble people, and I was more attracted to it. I am not just thinking of the outstanding few, the best, because somehow, I felt that it was often more important and more difficult to be good at something for a long time and quietly than to be the best at something for fifteen minutes once in a while. I was attracted to good people, and I felt something like this goodness in poetry, in the garden, and I feel it in beekeeping. Did Tolstoy also feel this goodness and beauty when he was beekeeping, and is that why he wrote so beautifully about bees? Perhaps.

The third aspect was more complex, because I thought – and this may be a naive idea – that tat if I started beekeeping more seriously at the age of forty, then perhaps, God willing, I could become a relatively acceptable beekeeper in my older years, and I could continue to do this beautifully alongside my writing, even in old age,. This cultivation, this culture, is what I thought could break down the external and internal barriers, heal the world and myself. I love the world of plants and animals, but it is also important to note that I have always loved detective stories and investigations, and beekeeping is often an investigation. This is certainly why the retired Sherlock Holmes took up beekeeping.

Incidentally, I also enjoy hexagonal, circle-based, complex computer strategy games, which are very similar to beekeeping with multiple colonies. I keep a diary of all my bee colonies, observe them, and summarize my experiences during the winter break. All this is very similar to scientific experimentation, where one experiments, observes, collects experiences, and only makes statements based on them. Another factor was that our home is located on the southeastern hillside, between Szentendre and Leányfalu, in the beautiful Danube Bend, at the meeting point of the Pilis and Visegrád mountains, which is a fortunate location for beekeeping.

In addition to all this, there was also an unexpected turn of events, so-called "fate," which played a role in my immediate immersion in beekeeping. The father of a longtime, good friend of mine was an experienced beekeeper, and this friend had called me earlier to ask me to visit him, but I kept putting it off. Then the COVID-19 pandemic took the old man away, and, as I wrote in another poem, his last words were, "I have fed the bees". I still have his hives and tools, and sometimes I even use them.

This quiet beekeeper was called József Miklósi, and thanks to the kindness of my friend and his brother, his bees and hives came to me, and suddenly I had seventeen families. I could talk at length about all this, but the point is that at the age of forty I found myself in the middle of a small apiary, with a lot of theoretical knowledge but very little practical experience, and I realized that Sándor Márai and Marcus Aurelius were not entirely right after all. Of course we don't know everything! There is a lot to learn, and practice is the best teacher.

Our hives are 24-frame, single and double family hives from Nagyboconád, which are only used in Hungary; stackable hives are more common around the world. I have read and learned a lot about beekeeping with these hives, but I am still learning. I love these hives because they are like books, with the frames built into them like the pages of a book, in which we write our golden verses, which is honey itself, together with the bees. So, I have a typical, small Hungarian apiary on the outskirts of Szentendre, and of course I read a lot of specialist literature, including the books of Barnabás Zsidei, who was a friend of my mentor and kept bees here in the neighboring village of Leányfalu. I am also very interested in new technologies. For example, we built a bee cabinet with lockable vents and a small ozone generator, which performed surprisingly well. Also, the solar wax melter I inherited from my brother Józsi works very well for me. I have added wheels to it so that during the summer I can constantly turn it towards the sun on the terrace, and it melts the used combs very nicely. In any case, I find it typical of beekeeping that there is always something to fix or repair. Perhaps I will not offend either the more professional or the more self-conscious beekeepers if I say that, for me, beekeeping is often like "adult Lego." A skilled beekeeper can be quite resourceful when it comes to solving previously unknown problems.

The author's home in Szentendre in 2023

In my first year of production, I constantly struggled with the health of my bees and tried to develop a sustainable mite protocol, but I still have room for improvement in this area. In the end, we had a very good honey harvest, but the two-month drought at the end of the summer and the resulting lack of pollen meant that climate change, human activity, deforestation, the disappearance of bee pastures, the use of chemicals, and invasive pests all have an impact on bees, making it increasingly difficult to keep them alive. I could say that bees are the most sensitive litmus test of environmental change; you can see directly from them how the planet is being destroyed.

I try to practice ethical beekeeping, which means that I always leave honey for the bees (yes, even acacia honey), I never completely empty the hive, and my primary concern is the well-being of the colony, meaning that true success is not measured by the amount of honey, but by the health of the bees. In addition to honey, we also collect pollen, make propolis solution, and my wife, Virág, who is my partner in this, and I use the wax melted in the solar wax melter to make Christmas candles. Beekeeping is not yet a matter of livelihood for us, but an opportunity for development, learning, and "rethinking our relationship with nature."

As I finish this book – in October 2023 – I am wintering with twenty-six families, and, thanks to my friends, I have also built a small apiary, or as they say elsewhere, a hive house, as a kind of laboratory, and I have begun to study the healing effects of bees: apitherapy. At this time of year, I always await spring eagerly.

I have made many mistakes and gained a lot of experience during my first years as a producer, and I try to draw strength from this and learn so that I can overcome the difficulties of the coming years. I am grateful to everyone who has helped me with this. I am grateful to my brother, my friends, my wife Virág, Péter Bross, Mihály Hudek, Mihály Hevesi, György Neumann, Krisztián Márton, Attila Miklósi, Imre Gazsó, and Andin Szűcs. Thank you all for your help. The truth is that without supporters, helpers, and good and useful advisors from whom one can and should learn, beekeeping is not such an easy thing.

I have read and researched a great deal about bees, not only as a beekeeper, but also as a philologist and poet. I wrote *The Art of Beekeeping*, which explores how important, exciting, and interesting the role of bees has been in human cultural history, literature, and art from the beginning to the present day, and what a special interaction has developed between bees and humans. My essays were published in the trade journal *Méhészet (Beekeeping)* over a period of two years, and now these essays have finally been compiled into a volume. A lot has been left out of this volume, such as a more detailed explanation of apitherapy or a description of the possibilities of chemical-free, practical beekeeping, but these would have stretched the already wide-ranging art of beekeeping too far.

I sincerely hope that my book will find its audience, whether they are beekeepers or not, people who love bees and nature and learn from them just as much as the author of these lines.

Weiner Sennyey Tibor's beehive

The Wandering Beekeeper

In memory of Anton Janša

Living together with our bees
and heading outdoors when called.
Observing shining fields,
finding good pastures for the bees.

Pollinating flowers,
collecting the abundant honey
taking only the surplus,
Letting the weak grow strong.

Observe, develop,
keep learning;
Look through the prism of the hive
To perceive the world's depths.

Wander in the homeland,
Carrying the flame of beeswax in the night
Bring the light of a pure spirit,
To open the resounding gate of the present.

We return to the bee house,
Turning work into honey,
writing sweet poems in sunlight
Beside the glowing steps.

Hibernating with our bees,
dreaming alertly, then, we must
understand the starry sky,
And so embrace our god.

Unije, August 10, 2021.

www.ingramcontent.com/pod-product-compliance
Ingram Content Group UK Ltd.
Pitfield, Milton Keynes, MK11 3LW, UK
UKHW061956041026
15591UKWH00016B/1028

9 781919 200491